Lamborghini's
HOLY TRINITY

To our grandparents, Tata and Patti, who cared so very much

Lamborghini's HOLY TRINITY

THE EVOLUTION OF THE COUNTACH, DIABLO AND MURCIELAGO BLOODLINE

THILLAINATHAN PATHMANATHAN AND ANNE CHRISTINA RECK

PEN & SWORD
TRANSPORT
AN IMPRINT OF PEN & SWORD BOOKS LTD.
YORKSHIRE – PHILADELPHIA

First published in Great Britain in 2026 by
Pen and Sword Transport
An imprint of
Pen & Sword Books Ltd
Yorkshire – Philadelphia

ISBN 978 1 03610 074 2

A CIP catalogue record for this book is available from the British Library.

Typeset in 12/16 Sabon LT Std by SJmagic DESIGN SERVICES, India.

Printed and bound in India by Replika Press Pvt. Ltd.

The Publisher's authorised representative in the EU for product safety is Authorised Rep Compliance Ltd., Ground Floor, 71 Lower Baggot Street, Dublin D02 P593, Ireland.
www.arccompliance.com

For a complete list of Pen & Sword titles please contact

PEN & SWORD BOOKS LIMITED
George House, Beevor Street, Off Pontefract Road, Hoyle Mill, Barnsley,
South Yorkshire, England, S71 1HN.
E-mail: enquiries@pen-and-sword.co.uk
website: www.pen-and-sword.co.uk

or

PEN AND SWORD BOOKS
1950 Lawrence Rd, Havertown, PA 19083, USA
E-mail: uspen-and-sword@casematepublishers.com
website: www.penandswordbooks.com

FSC www.fsc.org MIX Paper | Supporting responsible forestry FSC™ C016779

CONTENTS

AUTHOR BIOGRAPHIES

Dr Thillainathan Pathmanathan's fascination with Lamborghinis was spiked at the age of eight by the sight of the Marzal concept car. His interest in Sant'Agata's supercars has never waned since. Dr Pathmanathan's primary interest is in the wedge-shaped, longitudinal mid-engined, spaceframe-chassised, Bizzarrini-engined, Lamborghini flagships. He has authored the world's first stand-alone books on the Lamborghini Murcielago and the KTM X-BOW, and more recently has written books on the Lamborghini Countach and the Lamborghini Diablo. He has also authored the opening volumes of a dodecalogy of travel inspired socio-politico-economic essays. Dr Pathmanathan boarded at Cheltenham College, then obtained his medical degree from Pembroke College, University of Cambridge, with a degree in Physiology from the University of London en route. He trained as an eye surgeon in London, Southampton and Liverpool before undertaking a specialist Glaucoma Fellowship at the University of Toronto and was appointed a Consultant Eye Surgeon in 1999. He feels immensely privileged to hold, since 2001, joint temporary custodianship of Countach Chassis JLA12399, a totally factory-standard Rosso Siviglia 88½ 5000 QV – one of only fourteen such right-hand drive cars in the world. A sibling in the form of Murcielago Chassis 1564, an unaltered Arancio Atlas 2005 Roadster, arrived in 2014. His most precious automotive memories are of meeting and interviewing Paolo Stanzani, Giampaolo Dallara, Umberto Marchesi, and Marcello Gandini.

Dr Anne Christina Reck's first sighting of a guillotine-door Lamborghini V12 entranced and enchanted her. Her respect for these Sant'Agata flagships grew with each visit to the factory, and with every encounter with the people who conceived, developed, built and now preserve these automotive jewels. Dr Reck has co-authored books on the KTM X-BOW, the Lamborghini Countach, and the Lamborghini Diablo. She has also co-authored, edited, and acted as publishing agent for a dodecalogy of travel-inspired socio-politico-economic essays. Dr Reck obtained her medical degree from the University of Copenhagen, trained as an eye surgeon in London and Southampton, and did a specialist Retinal Fellowship at Moorfields Eye Hospital before being appointed a Consultant Eye Surgeon in 1999. Her most abiding car memories include the gala factory launch dinner of the Murcielago, watching the last Diablo roll off the Sant'Agata production line, and most of all setting eyes for the very first time on the spaceship-like Countach.

ACKNOWLEDGEMENTS

Some of the most important contributors have asked not to be named, but we thank everyone who has helped – this has been a team effort.

Foremost amongst these have been Paolo Stanzani, Marcello Gandini, Giampaolo Dallara, Umberto Marchesi, Horacio Pagani, Valentino Balboni and Loris Bicocchi, who were all kind enough to engage with us, and provide deep and absolutely invaluable insight into the Countach, Diablo and Lamborghini.

Maurizio Reggiani, Gerald Kahlke, Francesco Stevanin, Elena Ruosi, Chiara Sandoni and Dario Lucchese of Lamborghini S.p.A. and Rodrigo Ronconi of Polo Storico Lamborghini were very generous with their time, in giving us precious information, and letting us use various photographs, drawings, and handbook and brochure illustrations. Archivio Automobili Lamborghini and its consultant Barbara Grillini provided us with invaluable photographs, drawings, and information, as did Horacio Pagani and Giulia Roncarati of Pagani Automobili S.p.A.

Cav. Comm. Ing. Tonino Lamborghini, Ferruccio Lamborghini Junior and the Ferruccio Lamborghini Museo, Funo Di Argelato, Historical Archive very generously provided precious text, photographs and inside knowledge of the Lamborghini family. The Italian State, through its Ministry of Culture (Beni Culturali Italia) and Alberto Corteggiani, gave us priceless early Bertone/Gandini Countach drawings.

RM Sotheby's, in the form of its Executive Vice President and Head of Europe, Maarten ten Holder, and its Director of Research, Eugene Robertson, has been supremely generous in providing us with high-quality photographic portfolios of some of the most notable Countachs that have passed under the gavel of this world-class auction house. Thank you to its photographers Darin Schnabel, Nathan Leach-Proffer, Karissa Hosek, Teddy Pieper, Simon Clay, Erik Fuller, Juan Fernando Martinez Silva, Patrick Ernzen, Tom Gidden, Tim Scott and Dirk de Jager.

The National Motor Museum at Beaulieu, Hampshire, is the second home for many of our cars and we are deeply grateful to Jonathan Day, Tim David Wood, Russell Bowman and Stephen Munn for the use of the wonderful images taken in the Beaulieu photographic studio.

Ian Hunt has been a steadfast mentor and we have been very lucky to have had on board acknowledged Lamborghini experts like Stefano Pasini, Mike Pullen, Roberto Grimaldi, Sheikh Amari, Verna and Tadek Lipinski, Stephen and Beverly Ward, Alan Robb, and Bob and Veronica Forstner. We are very grateful to Rod Grainger of Veloce Publishing for having had the courage to give us our first break a few years ago. Taking on two enthusiastic but virginal authors was always a risk. Silvio Tschang and Giuseppina Maccioni, Ferruccio Lamborghini's secretary, welcomed us with equal warmth into La Fiorita, where Ferruccio chose to spend his last days. Jim Holder of *Autocar*, Colin Overland of *CAR* and Stuart Gallagher, Ryan Chambers and Nick Trott of *EVO* have been hugely kind and encouraging to two amateurs throughout this enterprise, and we would like to thank them for allowing us to use excerpts from their magazines. Mel Nichols, Peter Dron, Ian and Antony Fraser, Gautam Sen, Eiichi 'Eddie' Okado, Raymond Stofer, Marcel Wallenburg, Robert Hennessy, Nigel Ratledge, Richard Hawkins, Chris Ohlgren, Craig Nelson, Paolo Grasso, David Lillywhite, Jason Vidgoff, Steve Higgins, Robert Westall, Maurice Rizzuto, Peter Griffiths, Ryan Upward, Nick Elkan, Matt Bell (Pirelli Tyres UK), Chiara Guizzi (Fondazione Pirelli), Kirsty Pullen, Lucas Hutchings (Octane Collection), Simon Furlonger and Matthew Honeysett (Simon Furlonger Specialist Cars), Branko Radovinivic, Mark Sheldon-Lloyd, Sourein Jabourian, Eric Verdon-Roe, and Mark and Mary-Rose Hughes all helped enormously with information, contacts and photographs.

We are immensely grateful to Herman Dummer of The Gallery Aaldering, Gavin Halterman of We are Curated, Linus Otto and Pascal Stephan of Mechatronik, Thijs Verhage of Rock N Roll Classics BV, Nikolaj Jeppesen and Steffan Frisk of Selected Car Investment DK, and Jack Vass for their generosity in providing truly gorgeous photographs.

Finally, we must again acknowledge and thank those invaluable contributors who have asked to remain nameless. Some of your offerings were beyond priceless – your request for anonymity will always be respected.

INTRODUCTION

Lamborghini's Holy Trinity has long fascinated the aficionado, just as it has always intrigued the dispassionate. The Countach, Diablo, and Murcielago, are globally recognised motoring phenomena, and each has the inherent power to elicit some sort of emotional response from almost everyone, even the disinterested. This in itself is extraordinary and speaks volumes.

While whole libraries of books have been written about the Countach, only reams and quires of magazine and internet articles have been written about the Diablo and the Murcielago respectively. There must be a score or more of stand-alone books on the Countach dating back to Marchet & Coltrin's slim but authoritative 1981 treatise – which surely graces the bookshelves of every Lamborghini enthusiast. To date (2025) there have been far, far fewer stand-alone books on the Diablo and the Murcielago – in fact just two on the former, and only one on the latter. Stefano Pasini's book on the Diablo was written in 1991 and covers only the first of the eight Diablo variants and only the first eighteen months of the Diablo's eleven-year reign. Pathmanathan and Reck have written the other Diablo book and this is currently the only book that covers the full history and range of the Diablo. Pathmanathan has also penned the only current stand-alone book on the Murcielago.

But taken together all these books cover the essentials of the Countach, Diablo and Murcielago story. So what is the justification for yet another book on this trio – especially as your present authors have also already written a book on the Countach. Facts by definition remain true, precise, and unaltered, and there is little point in simply recycling information when many of the aforementioned books remain in print or are readily available on the second-hand market.

The reasons for writing this book are threefold, and we have wanted to pen it since about 2012, which is when we first seriously started thinking about providing our Rosso Siviglia 88 ½ Countach (Chassis JLA12399) with a V12-engined Sant'Agata sibling. The more we looked into the Diablo and the Murcielago the more we realised the following trio of justifications, amounting to a need, for a new Lamborghini book.

Firstly, the similarities between the Countach, Diablo, and Murcielago – cars seemingly different at first sight – were spectacularly striking, and to date no publication has attempted to draw the golden thread that weaves its way between these three vehicles.

Secondly, no publication to date has investigated and elaborated on the external factors that forced evolutionary change upon Sant'Agata – leaving the factory no choice but to transmute the Countach into the Diablo, and later, the Diablo into the Murcielago.

There were five key external forces that drove this evolution forwards – technological innovations, inter-manufacturer rivalry, and socio-politico-economic factors. Air pollution legislation, climate change pressures, driver, passenger and pedestrian safety laws, wars and revolutions, and economic crises and global recessions – all these had a truly profound impact on Lamborghini's engineers, designers, accountants, managers,

Jaguar Classic
LAT114M
E57 JAR

and ultimately its flagship supercars. In particular it is astonishing how Central Bank interest rates, and fuel availability and petrol prices, repeatedly threatened nothing less than Lamborghini's very survival in the period between 1971 and 2010 – and these need talking about. It is impossible to fully appreciate the Countach, Diablo, and Murcielago without knowing and understanding how all these external factors influenced Sant'Agata and the wider contemporary supercar world.

Thirdly, the Countach's position and status relative to its immediate predecessor flagship car, the Miura, needs reappraisal. The revolutionary Countach brought many more design and engineering innovations to the automotive world than the iconic Miura ever did, thus making the former the more important and the more instructive of the two. This book provides the objective evidence for this bold and novel statement.

Evolution is the key theme that will run throughout this book – simply because it was an unceasing series of innovations and developments that kept this Holy Trinity, and Lamborghini S.p.A. itself, alive over this thirty-nine year period. What greater evidence does one need of the constructive power of evolutionary-led meritocracy than that it kept the Countach, Diablo, Murcielago troika at the very top of the hotly-contested supercar pyramid for almost four decades. Equally, evolutionary forces were wholly responsible for this bloodline's death in 2010 – when external forces dictated that this triumvirate could no longer adapt to be the fittest, and therefore no longer deserved to survive.

Opposite: Lamborghini's Holy Trinity on display at the 2023 London Concours held in the grounds of the Honourable Artillery Company's Headquarters.

Right: Rear wheel arch of a 25th Anniversary Countach. Marcello Gandini's signature slashed and uprising rear wheel arch can be seen in varying forms in each of the Holy Trinity cars.

Rear wheel arch of a Diablo SV.

Rear wheel arch of a Murcielago LP 640.

First and Last, Front and Back – the first of Lamborghini's Holy Trinity peeks across the rear of the last.

CHAPTER 1

LAMBORGHINI'S HOLY TRINITY – THE CONTEXT

The Lamborghini Countach, Diablo, and Murcielago remain amongst the most venerated, sought after, and influential supercars of the late twentieth and the early twenty-first century. This threesome share the same design philosophy, the same drivetrain rudiments, the same external architecture, and the same ergonomic language. Together they almost define Lamborghini's early history.

The preceding Lamborghini Miura and the successor Lamborghini Aventador have fundamental drivetrain and chassis differences that make them entirely different beasts. At launch, neither of these cars were as revolutionary as the Countach. The Miura drew heavily from contemporary racing cars, and from the 1901 Autocar and the 1962 Bonnet Djet road cars, for its mid-engine layout, while its voluptuous body form was derivative and typical of the curvaceous track and sportscars of the 1950s and 1960s. Similarly, the Aventador offered little that was genuinely new – both its Audi-derived engine and its carbon fibre chassis had already been seen elsewhere in one form or another. And neither of these cars spawned a bloodline in the way that the Countach did.

The Countach spawned a dynasty that lasted almost four decades simply because Paolo Stanzani and Marcello Gandini designed a truly anarchistic car, with genuinely groundbreaking drivetrain technology and trailblazing styling, that could effortlessly stand the test of time. And because Sant'Agata's engineers and designers then worked relentlessly to develop and modify this germinal seed, allowing the Countach to evolve into the Diablo, and then into the Murcielago.

AUTOMOTIVE EVOLUTION

Lamborghini's Holy Trinity remains the exemplar of real-life automotive evolution. The Countach, Diablo, Murcielago bloodline started off in 1971, evolved continuously, and only died off in 2010. The overarching themes of this book are twofold: the conception and relentless technological development of these three supercars, and the socio-politico-economic factors that drove this evolution forwards.

The Cambridge Dictionary defines evolution as 'a gradual process of change and development', while the Oxford English Dictionary defines it as 'the process of developing into a different form'. Charles Darwin's seminal 1859 treatise *The Origin of Species by Means of Natural Selection* did not use the term evolution – this term only made an appearance in Darwin's 1871 book titled *The Descent of Man*. Similarly, the phrase 'Survival of the Fittest' which describes and explains the twin theories of Darwinian Evolution and Natural Selection was not coined by Darwin. Rather it was Alfred Russel Wallace who urged Darwin to adopt this phrase, and the philosopher Herbert Spencer who later further publicised this phrase in response to Darwin's groundbreaking work. The aforementioned term and phrase are both fundamental to understanding how Lamborghini's Holy Trinity survived for so long, and to comprehending why its three constituent supercars so completely dominated their contemporaries.

At its launch in 1971 the Countach LP 500 prototype immediately seized the supercar crown as it was uncontestably the most radical, most beautiful, most avant-garde, most accelerative, and fastest (or so the factory claimed) supercar, and therefore the fittest of the breed. The subsequent production Countachs managed to survive the arduous process of hanging on to this crown against ever more potent rivals by changing, developing, and adapting. Challenges materialised not only from competitor manufacturers like Ferrari, Maserati, Porsche, and Aston Martin, but also from government legislation on noxious emissions, occupant and pedestrian safety, fuel consumption, and tailpipe emissions. Distant socio-politico-economic factors like the fallout from far-off wars, faraway economic recessions, and academically-proven but often intangible man-made climate threats, were further impedances. When the Countach could evolve no more – principally because of its awful aerodynamics – it quickly became unfit, and almost immediately lost its right to survive in the supercar jungle.

And in response Sant'Agata developed, and in 1990 launched, the much more aerodynamic Lamborghini Diablo. And when in good time, and in the face of new and cutting-edge challengers entering the supercar gladiatorial arena, the Diablo was no longer the fittest, Lamborghini was forced to bow it out in favour of the Murcielago. There should be no surprise in these consecutive Sant'Agata ripostes – they were merely following what the proto-evolutionary advocates that preceded Darwin, and some of Darwin's closer contemporaries, had long advocated. Aristotle and Galenus, and Empedocles and the other Greek Atomists, had all subscribed to the ancient concept of sequential biological change. Similarly, Carolus Linnaeus, Comte de Buffon, Erasmus Darwin, Robert Grant, and Chevalier de Lamarck had all endorsed some or other form of species transformation. And in his *Philosophiae Naturalis Principia Mathematic*a of 1687 the polymath Sir Isaac Newton had proposed the concept of dynamic forces altering nature over time. In developing Lamborghini's Holy Trinity Sant'Agata was simply following in the footsteps of established scientific and philosophical doctrine and putting it into hard-nosed commercial practice.

EVOLUTIONARY DRIVERS

Supercars are the ultimate discretionary purchases, and this makes their manufacturers highly vulnerable to the vagaries of the global economy. The typical supercar buyer, and the average Lamborghini customer, is no different, is uniquely positioned to profit big when economic conditions are favourable, and is discordantly exposed to equally large losses when the markets dive. A recession here, a war there, a spike in oil price everywhere, compounded by social sentiments that change with the wind, often have an immediate and dramatic impact on supercar factory order books. This has direct repercussions on the funds available for future research and development, which in turn brings ramifications for the technical specification and the launch date of the next model or the next model variant.

An origami-sharp Rosso Siviglia 88½ Countach 5000 QV. Note the rear tyre wedge courtesy of a perennially unreliable handbrake.

A winged and more wind-cheating Diablo in the wilds of the Australian outback.

This sequence is very clearly seen on looking back at the history and evolution of Lamborghini's Holy Trinity. Ferruccio Lamborghini's eponymously-named supercar company was wrenched out of his hands by the Bolivian Economic Debacle, Rossetti and Leimer had to relinquish control of Sant'Agata after perennial cashflow problems, Chrysler under Iacocca relievedly offloaded its Italian jewel when it ran out of liquidity, and MegaTech, V Power, and MyCom were forced to sell up when the fallout from the Asian Financial Crisis made it impossible for them to balance the books. The development of the Countach and Diablo variants was regularly hamstrung, and the evolution of the Countach into the Diablo was repeatedly delayed, directly or indirectly, by factors as diverse as Middle Eastern wars, stock market crashes, Federal Reserve interest rate rises, realignments of political blocs, market pessimism, asset bubbles bursting, and ever more demanding government legislation brought on by environmental concerns and societal pressure.

But it was not all bad news – some of the Countach's, Diablo's, and Murcielago's most groundbreaking innovations and quantum-leaps came when Sant'Agata's was forced to respond to new and ever more stringent legislation, or through heightened customer expectations, or with the unveiling of a rival supercar boasting genuinely novel, state-of-the-art, technology. Lamborghini's periods of advancement and periods of decline between 1971 and 2010, and the Holy Trinity's evolution, can only be understood when refracted into its constituent parts and viewed through this socio-politico-economic prism.

Opposite: Aerodynamics dictated the shape of the Murcielago's bodyshell, but decapitating this Arancio Atlas Roadster compromised its drag coefficient.

A DIARY OF THE RECESSIONS, WARS, AND SOCIO-POLITICO-ECONOMIC EVENTS THAT DROVE THE HOLY TRINITY'S EVOLUTION

1798	William Wordsworth's Romantic Movement identifies mankind as a threat to nature. This is Modern Europe's first introduction to proto-environmental concerns and to pollution anxiety.
1896	Svante Arrhenius, the father of anthropogenic climate change, publishes his seminal paper establishing a parallel between the Earth's increasing surface temperature and its increasing atmospheric carbon dioxide levels, and postulates a man-made causal link between the two (the Greenhouse Effect).
1967	The Six-Day War between Israel and a coalition of Arab states leads to the 1967 Oil Embargo.
1970	Public Health Cigarette Smoking Act signed by Richard Nixon recognises the damaging effects of noxious gases.
1971	The Concept Lamborghini Countach LP 500 Is Unveiled.
1971	The Nixon Shock. The United States cancels dollar to gold convertibility, freezes wages, and imposes tariffs on imports.
1972	The Bolivian Financial Crisis forces its government to renege on a contract for 5,000 Lamborghini tractors, which ultimately forces Ferruccio Lamborghini to sell his supercar company.
1973	The Yom Kippur War, precipitates the 1st Oil Crisis, and triggers the 55mph speed limit in the United States – just as the 196mph production Countach is about to be launched.

A disordered universe – a white Murcielago LP 640 interrupts the natural sequence of a blue Diablo VT following on a black 25th Anniversary Countach.

Precious cargo – An LP 640 Murcielago atop a Diablo SV.

1973–75 Global recession caused by a combination of the 1973 oil crisis, the Vietnam War deficit, and the collapse of the Bretton Woods Agreement subsequent to the Nixon Shock.

1974 The Production Lamborghini Countach LP 400 Is Launched.

1974 World population reaches 4 billion.

1975 Altair 8800, the first commercially successful personal computer, is released. Electronic control units would increasingly be used in the Holy Trinity.

1979 The Iranian Revolution ends on the 11th of February, and Shah Reza Pahlavi, a prolific Lamborghini supercar collector, is overthrown.

1979 The Iran Hostage crisis begins on the 4th of November.

1979 The 2nd Oil Crisis.

1980 The Iran-Iraq War begins in September.

1981 Global recession caused by oil prices spiking following the Iranian Revolution, as well as by the tight monetary policy introduced by the Federal Reserve to control inflation.

1986 Japanese asset bubble collapses. Japan is by now a key Lamborghini market.

1987 Black Monday Stock Market Crash caused by rising interest rates, overvalued stocks, and the persistent and growing United States trade and budget deficits.

1987 World population reaches 5 billion.

1989 The Revolutions of 1989 bring down communist and authoritarian regimes around the world, exemplified by the Fall of the Berlin Wall, the Collapse of the Soviet bloc, and the Tiananmen Square Protests. China, Russia, and some former USSR satellites would soon become lucrative markets for Lamborghini.

1990 The Lamborghini Diablo Is Launched.

1990 Iraq's Saddam Hussein invades Kuwait.

1990 The Intergovernmental Panel on Climate Change finally acknowledges that rising atmospheric carbon dioxide levels and rising global temperatures are causally linked to human activity.

1990 Global recession triggered by the Federal Reserve raising interest rates to tame inflation, and also by high oil prices, debt accumulation, and consumer pessimism.

1991 1st Gulf War: the Iraqi army is expelled from Kuwait.

1992 World Oceans Day celebrated in conjunction with the Earth Summit in Rio de Janeiro. Hurricane Andrew, which kills sixty-five people and causes US$ 26.5 billion in damages, is linked to man-made climate change.

1994 The United States invades Haiti to restore President Aristide to power. One of Aristide's predecessors, Francois 'Papa Doc' Duvalier, owned a Miura and an Espada and his close associate Albert Silvera owned a Miura SVJ and a Countach LP400. Jean-Claude 'Baby Doc' Duvalier was equally keen on supercars.

1995 World Trade Organisation established.

1997 Asian Financial Crisis indirectly leads to Lamborghini being sold to the VW-Audi conglomerate.

1999 World population reaches 6 billion.

2001 The Lamborghini Murcielago Is Launched.

2001 September 11th attacks on New York's Twin Towers. The United States invades Afghanistan and temporarily topples the Taliban when it refuses to hand over Osama bin Laden.

2001 Global recession following the September 11 attacks, the collapse of the speculative dot-com Bubble, and the Argentinian Economic Crisis.

A bevy of wedge-shaped, Bizzarrini-engined, Holy Trinity beauties.

Before the Murcielago – a parade of Miuras, Countachs, Diablos, Urracos and Espadas.

2003 2nd Gulf War: The United States invades Iraq and topples Saddam Hussein under the pretext of Iraq's non-existent Weapons of Mass Destruction. Saddam's son and one-time heir apparent, the notorious Uday Hussein, was a prolific supercar collector who owned several Lamborghinis including a Countach, a Diablo, and an LM002.

2005 Kyoto Protocol comes into effect. It sets binding emission reduction targets for thirty-seven countries. It was superseded by the Paris Agreement in 2016.

2007 The Great Recession triggered by the subprime mortgage crisis. Collapse of the very largest of financial institutions including Fannie May, Freddie Mac, Bear Sterns, and Lehman Brothers leads to an international banking crisis with serious fallout for the automotive industry.

2008 Tesla Roadster launched – the first mass produced lithium-ion battery electric car represents an existential challenge to internal combustion engine cars.

2008 Stock markets plunge as the Great Recession bites.

2009 Bitcoin launched. Millennial cryptocurrency traders would soon become a new and key demographic in Lamborghini's ever-younger customer base.

2009 The European Debt Crisis.

2009 The BRICS Economic Bloc established – Russia, India and China would soon become important Lamborghini markets.

2010 World population reaches 7 billion.

2010 The last Murcielago rolls off the production line and Lamborghini's Holy Trinity bloodline becomes extinct.

ROAD SAFETY AND ENVIRONMENTAL LEGISLATION THAT IMPACTED LAMBORGHINI'S HOLY TRINITY, AND SANT'AGATA'S RESPONSE

The world's first recorded fatal road traffic accident took place on 31 August 1869 when Mary Ward, an Irish scientist, fell out of her cousin's steam car and died after it ran over her. In 2023 road traffic accidents were responsible for approximately 1.2 million deaths worldwide, and somewhere between 20 and 50 million non-fatal injuries worldwide. In the United States a pedestrian is injured by a motor vehicle every eight minutes.

Road safety advocacy and legislation, spurred on especially by Ralph Nader's seminal 1965 book *Unsafe At Any Speed*, dove-tailed neatly with the nascent environmental activism and anti-pollution campaigns of the 1960s, 1970s, and 1980s. Below is a brief list of some of the notable legislation that impacted on the contemporary automotive industry and influenced the evolution of the Countach into the Diablo and thence the Murcielago. Note that although the Lamborghini factory was not officially exporting any cars to the United States until the later stages of the Countach's production cycle, it still had to bend its knee before this highly influential country's (and to European and Japanese) regulatory institutions when developing its supercars.

LEGISLATION

United States National Highway Traffic Administration's road safety legislation

United States Department of Transport's driver, passenger, and pedestrian safety legislation

United States Clean Air Act legislation

United States Environmental Protection Agency legislation

United States's Fuel efficiency targets and the blanket 55mph speed limit – justified on grounds of national security (through fuel security) and air pollution

EURO standards for Vehicles

European New Car Assessment Programme

European Union Ambient Air Quality Directives

European Union Fuel Quality Directives

Japanese National Safety and Environment Laboratory legislation

Japanese Ministry of the Environment Emissions and Fuel Efficiency legislation

THE NEW TECHNOLOGY THAT SANT'AGATA PROGRESSIVELY INTRODUCED IN ANSWER TO EVER MORE STRINGENT LEGISLATION

Front and rear bumpers

Wing mirrors

Fixed faired in headlights in place of retractable headlights

Pre-tensioner seatbelts
Fuel injection in place of carburettors
Driver and passenger airbags
Anti-brake locking (ABS) system
Dynamic Brake Rear Proportioning
Carbon-Ceramic brakes
Revival of dry-sump lubrication
Variable valve timing on both the intake and exhaust valves
Advanced ECU's regulating the three-way catalytic convertors and the oxygen sensor system
Chassis deformation and Crumple zones
Permanent all-wheel drive instead of two-wheel drive
Traction control system
E-gear robotised manual transmission with Corsa and Low Adherence modes
Bi-Xenon and LED lighting
Weight saving through the use of composites, particularly carbon fibre
Improved active and passive safety by increasing chassis torsional rigidity with supplementary ultra-high strength sectional steel and carbon fibre struts
Active Roll-Over Protection System for the Roadster variants
Active, electronically controlled shock absorbers
Active, real-time, aerodynamic optimisation using two motorised, electronically controlled, movable flank air scoops for the Variable Airflow Cooling System
Active, real-time, aerodynamic optimisation using a motorised, electronically controlled, retractable rear spoiler

CHAPTER 2

GENESIS OF THE COUNTACH – THE HOLY TRINITY'S PROGENITOR

The Countach was a revolutionary car. The Diablo and Murcielago were evolutionary variants of the Countach. These latter two cars should bear no shame for this. The Countach was after all a seminal supercar, arguably the most influential of the modern era – both from the aesthetic viewpoint and from the engineering standpoint. If the genius of Sant'Agata's designers and engineers allowed them to conceive the Countach, it was their dogged perseverance that enabled them to mastermind the evolution of the Countach into the Diablo and thence the Murcielago – thereby making these successor cars fit for purpose going into the last decade of the last century and the first decade of the new millennium – despite ever more stringent and demanding legislation, customer expectations, and inter-manufacturer rivalry.

FERRUCCIO LAMBORGHINI

At its core the Countach adhered strictly to the singular diktat that Ferruccio Lamborghini laid down for his eponymously named supercar company in 1963 – that it would only build Gran Turismo cars. In other words, road cars. Ferruccio's Automobili Lamborghini S.p.A. would not be in the business of building racing or quasi-racing cars. Ferruccio was a shrewd and battle-hardened businessman, and following his infamous, well-documented, and ill-tempered encounter with Enzo Ferrari, Ferruccio declared that his new automotive venture would build 'the perfect car'– by which he meant cars that would eclipse Ferrari's road cars. Ferruccio was too astute and too focussed to be tricked into competing with Enzo on track. The Countach, Diablo, and Murcielago were all designed to be road cars, albeit the most superlative Gran Turismo cars of their respective generations.

THE LAMBORGHINI MIURA

The Countach came into being, as did the Diablo and the Murcielago, through need. The Lamborghini Miura, the Countach's spectacular predecessor, suffered from fundamental design flaws – principal amongst which were high-speed front-end lift, insufficient chassis rigidity, and a tendency to snap oversteer. Sant'Agata's designers, engineers, and test and development drivers were unable to correct these faults despite two further iterations (the S and SV variants) over a seven-year period (1966 to 1973). Furthermore, by the end of the Miura's production run Lamborghini's rich and never-quite-satisfied customers had grown accustomed to the Miura's aesthetics and their jaded palates were now crying out for a tasty new morsel. Additionally, Lamborghini S.p.A. was thriving, and following the debut of the 350GT, Marzal, Miura, and Espada, had established a reputation as the most avant-garde supercar manufacturer. It now needed inspirational new blood in the form of a state-of-the-art flagship if it was to hang on to this hard-earned and highly treasured crown. By early 1970 it was clear that the Miura was no longer fit to compete at the highest levels, and therefore no longer deserved to survive. Automotive evolutionary forces were at work here, and the iconic Miura was dethroned by the spectacular Countach.

THE GOLDEN THREAD

The highly successful Countach boasted the following primary features and innovations that would in time become defining features of Lamborghini's Holy Trinity: a hand-welded spaceframe chassis, a Bizzarrini V12 engine, a longitudinal engine orientation, a South-North engine gearbox layout, an extreme wedge shape, and guillotine doors that rose upwards towards the heavens. Secondary features that featured in the earlier cars but were dropped as the bloodline evolved include retractable headlights and hand-beaten aluminium body panels.

THE COUNTACH LP 500 PROTOTYPE

The Countach LP 500 prototype was first shown in the huge Palexpo Halls at the 41e Salon Auto Geneve on 11 March 1971. This striking jewel, the runaway star of that year's motor show, actually made its debut on the Bertone stall and not the Lamborghini stand. Flocks, with mouths wide agape, crowded around the Grugliasco-based company's tiny booth, stunned by the LP 500's bright yellow colour, extreme wedge shape, guillotine doors, and retractable headlights. What these onlookers couldn't see, but which Lamborghini and Bertone proudly and loudly proclaimed, was the Countach's revolutionary drivetrain layout, which was designed to correct all of the Miura's dynamic flaws. Orders gushed in, despite neither Lamborghini nor Bertone giving any indication that this prototype Countach would ever make it into production. Blank cheques were handed over despite neither company giving the slightest hint of how much a production Countach would cost. It would be fair to say that the LP 500 had, within minutes of its unveiling and without turning a single wheel, proven itself able to stand toe-to-toe with the fittest of contemporary supercars, and so fully deserved to survive and make it into production.

THE LP 500 PROTOTYPE'S REVOLUTIONARY FEATURES

By Darwinian standards the LP 500 more than deserved to survive because it not only overshadowed every one of its rivals, but because it also brought into play new design and engineering solutions in answer to pressing automotive problems. Many of these solutions have stood the test of time – some remaining relevant through to the end of the Murcielago's production run in 2010, while others remain in play even today, albeit in slightly altered form – thus proving that revolutionary approaches can have genuine engineering worth, over and above simply shocking Motor Show attendees.

Because these revolutionary features are so critical and fundamental to the Countach, Diablo, and Murcielago, they are summarised here now, before being examined in detail later. These revolutionary features, first seen on the LP 500 prototype, are best divided into design features and drivetrain features. For completion the key engineering features that were carried over from earlier Lamborghini models into the Countach and thence into the Diablo and the Murcielago are also summarised below.

Opposite: The authors' Holy Trinity obsession was triggered by this Autocar magazine's front cover featuring the Countach LP 500 prototype.

DESIGN FEATURES

Extreme wedge shape
Guillotine doors
Retractable headlights
Kammback tail
Trapezoidal design motif
Slashed and uprising rear wheelarch
Slatted cooling vents
Futuristic cockpit and dashboard layout
Banana-shaped seats
Periscope rearview mirror – which sadly never made it into production
Hand-beaten aluminium body panels employed in Superleggera-style construction – later supplanted by composites
Use of exotic materials – Avional, Glaverbel, Magnesium – later supplanted by Nomex, Kevlar and Carbon fibre

Above: The father of the Countach, Paolo Stanzani, at the Murcielago launch in 2001.

Opposite: Stanzani's revolutionary South-North engine-gearbox orientation was carried through to the Diablo and the Murcielago.

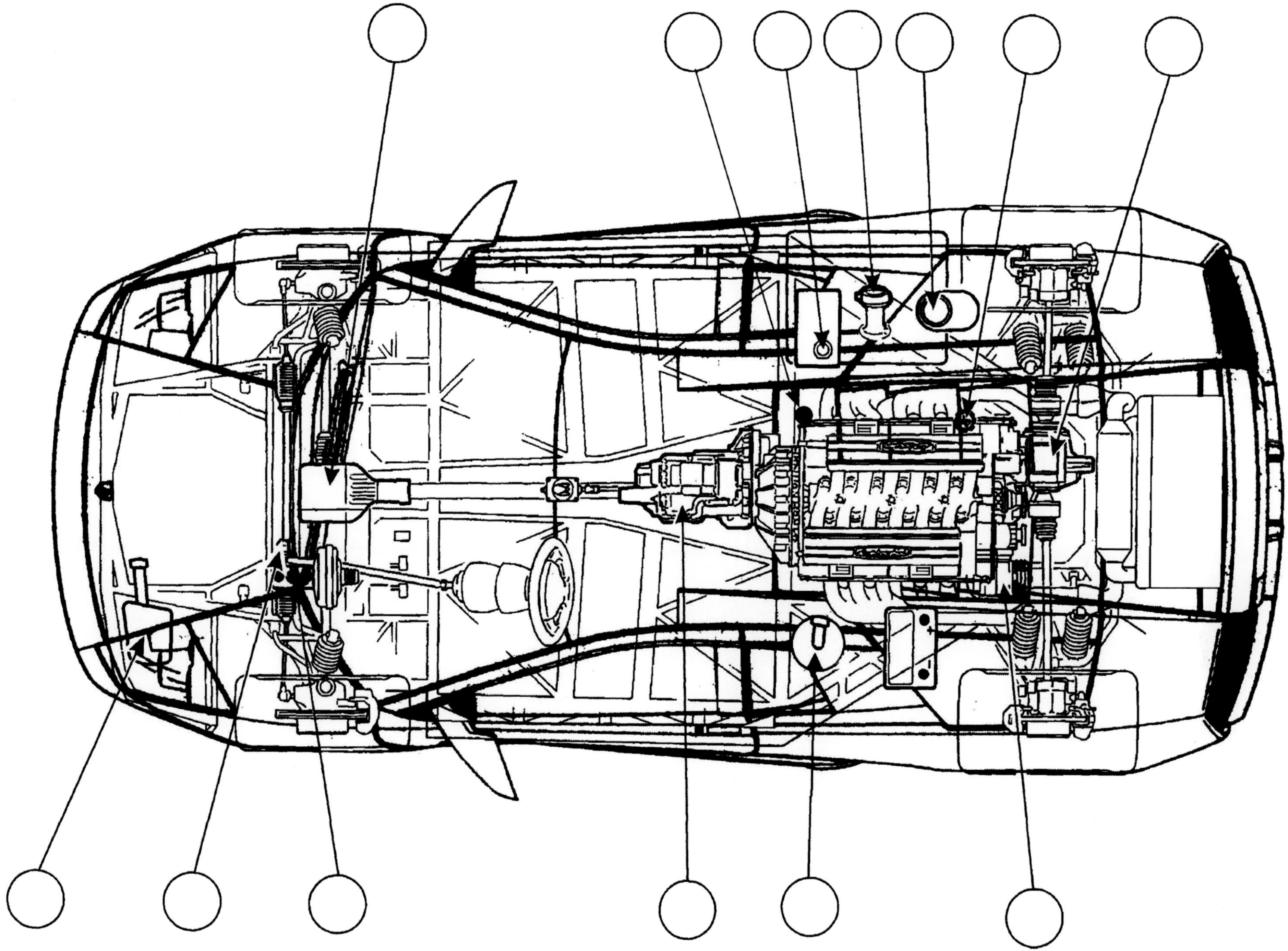

CHASSIS AND DRIVETRAIN FEATURES

The LP 500 prototype's semi-monocoque chassis would quickly evolve into a hand-welded tubular spaceframe chassis.
Longitudinal orientation of the Bizzarrini V12 engine
Paolo Stanzani's truly brilliant and genuinely revolutionary South-North engine-gearbox layout
Use of exotic materials – Elekton

KEY CARRY-ON FEATURES FROM EARLIER LAMBORGHINI MODELS

Bizzarrini V12 engine
Rear/Mid-engine layout
Rear wheel drive
Cooling innovations to address Lamborghini's perennial problems with the Bizzarrini V12 engine's prodigious heat output

DREAMING THE IMPOSSIBLE AND BRINGING IT TO LIFE – THE MEN RESPONSIBLE

Two men, Paolo Stanzani and Marcello Gandini, were responsible for dreaming up the Countach concept, and then working relentlessly and with intelligence to bring it to production. But both these two young men – they were still in their early thirties when the Countach concept was being developed in 1970 – could never have succeeded so completely without Ferruccio Lamborghini and Nuccio Bertone trusting them implicitly and giving them free licence to pursue their wildest dreams.

And thereafter the Countach's gestation was neither easy nor straightforward. Happily, talented and enthusiastic help was at hand. Stanzani had as his consiglieri Massimo Parenti, a meticulous engineer who has never received the full credit he justly deserves for his hard graft in taking the LP 500 from motor show concept car to production reality. Stanzani's other right-hand man was Bob Wallace, Lamborghini's legendary test driver, who patiently and methodically worked through each of the many challenges that the LP 500 prototype car threw up. And once the Countach was in production, it required the exceptional talents of Giampaolo Dallara and Giulio Alfieri, both legendary engineers, to progressively evolve the drivetrain rudiments so that the Countach could prevail at the very top of the supercar tree for an almost unbelievable seventeen years (1973 to 1990). For comparison, bear in mind that many of today's supercars have a lifespan of barely six years. Valentino Balboni, whose name is almost synonymous with that of Lamborghini S.p.A. itself, his understudy Loris Bicocchi, and multiple-time World Rally Champion Sandro Munari, honed the driving dynamics of the many Countach variants, test-driving them at high speed over long distances, and thereby ensuring that the Countach's acceleration, top speed, braking, and cornering abilities were not embarrassingly eclipsed by its stunning aesthetics. For secrecy against the prying eyes of rival supercar manufacturers and the motoring press, these test drives were often conducted in the depth of night, ignoring the additional dangers involved in such high-speed, witching-hour jinxes. And Horacio Pagani – of Zonda and Huayra fame – redesigned the 25th Anniversary Countach's bodywork. If Pagani's reworking of Marcello Gandini's masterpiece was not met with universal acclaim, his work on the advanced composite materials used in the Countach Evoluzione mobile test platform was widely recognised as outstanding.

Drive from the Countach's V12 Bizzarrini engine is first carried vanward to the forward placed gearbox before being sent aft to the rear differential via a set of drop-gears and an in-sump driveshaft.

The knowledge gained through the Countach Evoluzione would go on to inform the design of the 88 ½ QV and 25th Anniversary Countachs, and all the subsequent Diablo and Murcielago variants. Part of the Countach's charm and mythology lies in the very special people who conceived, developed, and modified it.

PAOLO STANZANI AND THE COUNTACH'S REVOLUTIONARY DRIVETRAIN

It is difficult to overstate the importance and significance of Paolo Stanzani's South-North engine-gearbox layout. This was a completely new concept that had never previously been expressed aloud in public, let alone been put into series production. The idea was entirely novel, and it succinctly answered several important drivetrain challenges. But it was a complex affair getting this engineering concept off the drawing board and into a Countach chassis.

That the very first prototype of this genuinely revolutionary South-North engine-gearbox layout worked almost perfectly straight out of the box speaks volumes about Paolo Stanzani's attention to detail and engineering expertise.

But this South-North engine-gearbox layout was only one part, albeit the most innovative part, of a greater whole designed to ensure that the Countach would be the most dynamic and most avant-garde of its supercar contemporaries.

The Countach's predecessor's rear/mid-engine layout had already proven itself – in terms of packaging, performance and pedigree – and had to be retained. Lamborghini's well-heeled customers loved the Miura's race-car like rear/mid-engine layout, and in particular its association with Formula 1, and this was therefore a key consideration that Ferruccio Lamborghini and Nuccio Bertone had to bear in mind when deciding on the brief that these two mature business owners would hand down to their Young Turks – Paolo Stanzani and Marcello Gandini. Having said which, both Stanzani and Gandini would probably have chosen this layout anyway, had they been given a free hand.

Why? Because there were good design and engineering reasons for choosing the rear/mid-engine layout, just as there were very good technical reasons for choosing a rear-wheel drive configuration and a longitudinal orientation of the Bizzarrini V12 engine. And, of course, these technical justifications remained just as relevant to the Diablo and the Murcielago as they were to the Countach concept.

A mid-engine layout places the car's engine and gearbox – the heaviest singular components of a supercar – close to the vehicle's centre, and therefore close to the car's centre of gravity and its rotational axis. This confers an exceptionally low moment of inertia upon the vehicle, which in turn affords the car exceptional agility and manoeuvrability. Another key reason why Stanzani and Gandini would probably have chosen a mid-engine layout anyway is because this layout readily allows the designers to approach a 50/50 front/rear weight distribution. With a 50/50 weight distribution all four tyres are forced into working equally hard – making the car more balanced and neutral when cornering, and helping with traction when accelerating and braking.

Similarly, combining a rear/mid-engine layout with a rear-wheel drive configuration is highly desirable because significant mass is placed over the driven tyres. This means that the rear driven tyres have additional traction when accelerating, and are also better able to assist the front tyres under hard braking. The added traction that this combined layout affords the rear tyres also decreases the likelihood of a rear wheel locking up and provoking a spin.

A Countach's massive drivetrain being removed for a wallet-draining clutch replacement.

A Murcielago Roadster's drivetrain being removed for a only slightly less wallet-draining rear main seal replacement.

There are two additional advantages of combining rear-wheel drive with a rear/mid-engine layout. Firstly, this combination eliminates engine torque transfer to the chassis. Because the engine can be attached directly to the gearbox, the torque can flow directly from the engine to the gearbox, rather than requiring the torque to flow from one end of the car to the opposite end, which inevitably brings along an element of chassis flex that will degrade handling. The second advantage conferred by this combination is that the free space liberated at the front of the car where the engine would normally reside can now be used as a frontal safety crush zone instead.

But there is one major dynamic disadvantage to employing a combined rear-wheel drive and rear/mid-engine layout. Putting all the major masses so close to one another, and so close to the car's centre of gravity and centre of rotation means that a more unannounced, more sudden, more violent, and more uncatchable spin is more likely to occur. This was one of the Miura's key failings, and as talented as Giampaolo Dallara was he was unable to engineer basic physics out of the Miura.

And this is where and how Paolo Stanzani's novel and brilliant idea of a South-North engine gearbox orientation, combined with a longitudinal placement of the stalwart Bizzarrini V12 engine, immediately takes the Countach to an altogether higher dynamic plane than the Miura could ever dream of climbing to.

Paolo Stanzani was no more able to defy physics than Giampaolo Dallara, but he had one major advantage. Stanzani's Countach was a clean-sheet design with no preconditions or preconceptions, and Ferruccio Lamborghini was happy for Stanzani to employ as much blue-sky thinking as he could muster.

So Stanzani chose to orientate the Bizzarrini V12 engine longitudinally – essentially turning it through 90 degrees compared to its orientation in the Miura. By distributing the engine's mass more equally along the full length of the Countach, rather than having it concentrated in the centre a la the transverse-engined Miura, the drivetrain mass was now closer to the Countach's centre of aerodynamic pressure as well as to its centre of gravity. This more forgiving weight distribution significantly reduced the risk of high-speed, front-end lift and snap oversteer – both serious Miura foibles that its rich but not necessarily talented owners were largely incapable of coping with. Stanzani had long determined to tame if not eliminate these twin evils.

This longitudinal V12 engine distribution did however mean that the Countach suffered from a higher moment of inertia than the Miura. But Stanzani, Gandini, Bertone, and Ferruccio Lamborghini were all in agreement that the resultant small loss of agility was a perfectly acceptable price to pay for better roadholding and safer handling. Furthermore, the longitudinal engine orientation allowed for better engine cooling, and for easier access to the engine and its ancillary parts for routine servicing, while also reducing engine and gearbox noise and vibration intrusion into the Countach's cabin.

But the longitudinal installation of a massive V12 engine presents a major disadvantage in that it takes up so much space lengthwise, especially once a hefty gearbox capable of taking all that torque, and a rear differential, and a final drive, are all attached to one end of the engine. And this is where Paolo Stanzani's ingenious South-North engine-gearbox orientation rides to the rescue.

This South-North engine-gearbox orientation involves turning the conventional engine-gearbox layout through 180 degrees and attaching the gearbox to the front of the engine. The torque from the engine is first directed forwards into the gearbox, from where it is directed downwards by a drop-down gearset, and then directed backwards through a sealed lubrication chamber within the engine sump, to the rear differential and the final drive.

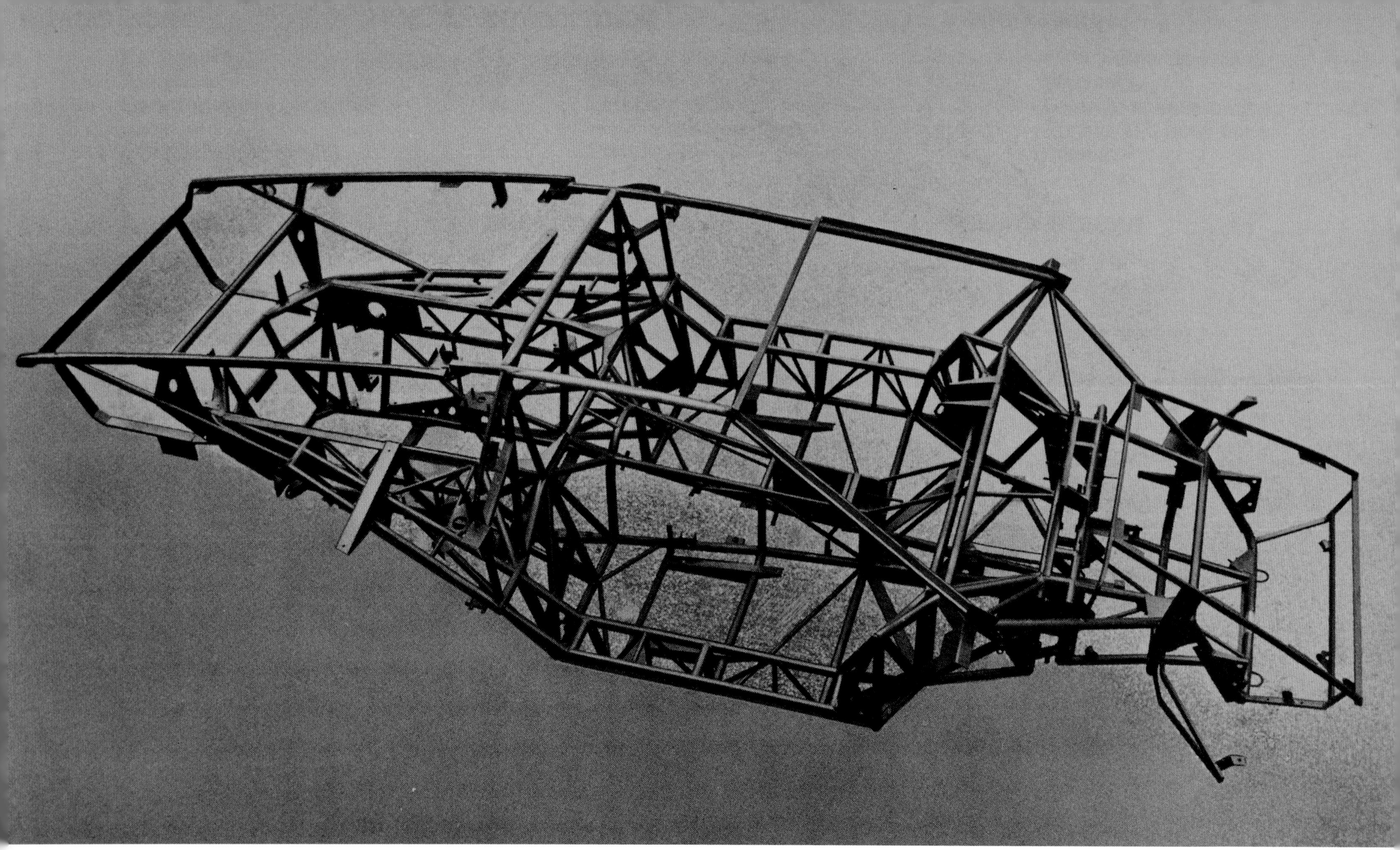

The Countach's intricate and definitive hand-welded spaceframe chassis was constructed from round-section steel tubes by Marchesi & C. Diablo and Murcielago chassis were built using less romantic, more prosaic and more functional square-section tubes as these were easier and quicker to weld, making these later spaceframes much cheaper to construct.

This arrangement markedly shortens the total length of the drivetrain, allows for better weight distribution along the whole length of the vehicle, dispenses with the need for lengthy and cumbersome gear cables and gear rods as the cockpit gear lever now sits immediately above the gearbox, and moves more of the overall drivetrain mass further forwards, which in turn allows the front tyres to do more work during acceleration, braking, and cornering manoeuvres.

Employing this unique drivetrain arrangement meant that the Countach was shorter in both wheelbase and overall length than the Miura, despite orientating the colossal Bizzarrini V12 longitudinally rather than transversely as in the earlier car.

There were four disadvantages to this South-North engine-gearbox orientation, but Stanzani felt that these were easily outweighed by the advantages. Firstly, the need to accommodate the driveshaft within the engine sump raised the overall height of the engine and so the car's centre of gravity, which was detrimental to the car's handling. Secondly and thirdly, the modifications needed to accommodate the sealed driveshaft within the sump increased the weight of the car and the cost of manufacturing and assembling the drivetrain. Finally, the forward location of the clutch meant that the whole engine-gearbox complex would have to be removed from the engine bay every time the clutch had to be replaced – a complex, labour-intensive, fraught-with-risk, and costly procedure. The last-mentioned disadvantage is why Countach, Diablo, and Murcielago owners shudder, almost faint, and reach for the smelling salts at the very mention of a clutch replacement!

Every Countach variant and several Diablo variants employed a rear/mid-engine configuration combined with rear-wheel drive. It was only with the introduction of the Diablo VT and its attendant four-wheel drive technology that Sant'Agata departed from this tried and tested arrangement. All the Countach, Diablo and Murcielago variants deployed their Bizzarrini V12 engines longitudinally, and all variants used an iteration of Stanzani's South-North engine-gearbox orientation – obviously with the necessary modifications for those later Diablo variants and all the Murcielago variants (bar the two-wheel drive Murcielago R-GT) which boasted a four-wheel drive system.

GIOTTO BIZZARRINI AND LAMBORGHINI'S STALWART V12 ENGINE

Every Countach, Diablo and Murcielago has at its heart a Bizzarrini V12 engine. Every Sant'Agata flagship from the very first Lamborghini 350 GTV prototype (1963) to the very last Lamborghini Murcielago (2010) has employed this same basic engine – albeit altered through a never-ending series of modifications and refinements. Lamborghini's Bizzarrini V12 engine evolved with each model change, and often with each model variant change. Nothing, not even the guillotine doors or the hand-welded spaceframe chassis, cries out the Lamborghini name quite as loudly as the Bizzarrini V12 engine does.

And yet this most emblematic of engines had an unbelievably difficult conception, gestation, and infancy. Enzo Ferrari rejected it, Ferruccio Lamborghini abhorred the power and torque delivery of its first iteration, and Giampaolo Dallara and Paolo Stanzani had to sweat hard labour to make it fit for purpose as the powertrain for a line of exclusive and hyper-expensive Gran Turismo road cars.

The story of this stalwart engine begins well before Lamborghini S.p.A. was even set up. Giotto Bizzarrini was a highly talented engineer who qualified with top honours from the University of Pisa before starting work at Alfa Romeo, where he befriended the legendary Carlo Chiti. Shortly thereafter Chiti was appointed Chief Engineer at Ferrari,

Umberto Marchesi at his factory in Via dei Tipografi, Modena – the birthplace of every production Countach's skeleton.

Marcello Gandini with the authors in September 2023. Gandini designed the Countach and the Diablo, and Luc Donckerwolke was honest enough to say that he got inspiration, and took styling cues, from the Maestro's design language when fashioning the Murcielago's haut couture.

and Chiti took Bizzarrini along with him to Maranello. Both Chiti and Bizzarrini fell victim to Ferrari's infamous 'Palace Revolution' in the autumn of 1961, but before he was forced to leave Bizzarrini designed a 1.5 litre V12 engine for Enzo Ferrari's Formula 1 race cars. However, Enzo had other engine options available and Bizzarrini's design never made it off the drawing board.

When Ferruccio Lamborghini decided to make 'the perfect Gran Turismo car', he knew that nothing less than a V12 engine would do. Countless luxury car makers had long rejected Henry Ford's diktat that 'A car should not have any more cylinders than a cow has teats', and by the early 1960s numerous six-, eight-, twelve-, and even sixteen-cylinder engines had found their way into road and race cars.

The twelve-cylinder engine held a particular and totally understandable attraction for Ferruccio Lamborghini. Twelve-cylinder engines have an inherent mechanical balance and they also scream out luxury and exclusivity. Not for nothing did Phil Hill, the 1961 Formula 1 World Champion, lavish praise upon his multiple race-winning Ferrari V12 engines with the following words: 'beautiful flexibility', 'multiple stages of performance', 'the more revs you use the more torque and power are available'. The Berlin Philharmonic's long-standing principal conductor, Herbert von Karajan, was no less lavish in his ovation of Enzo Ferrari's road-going V12 engines, saying, 'When I hear your twelve cylinders, I hear a burst of harmony that no conductor could ever recreate.' And Damon Hill, the 1996 Formula 1 World Champion, spoke of the V12 engine's pulsation pattern as 'the sonic equivalent of strawberry mousse and cream'. No, if Ferruccio Lamborghini was to equal if not better his nemesis Enzo Ferrari, then nothing less than a V12 engine would do.

When in the very early 1960s Ferruccio Lamborghini was thinking about setting up his Gran Turismo manufacturing company, he entered into discussion with Neri & Bonacini, a Modenese coachbuilding company, whose directors casually mentioned that Giotto Bizzarrini was an excellent engine designer. Bizzarrini set up Societa Autostar, and through this company received the commission from Ferruccio to design, develop, and test a V12 engine for Automobili Lamborghini S.p.A.. The commission stipulated a minimum power output of 350bhp, with an incremental bonus for exceeding this output and an incremental deduction for undershooting it. Half of the 4,500,000 Lira fee was paid immediately after both parties had signed the contract, with the remaining half to be paid on delivery of the tested prototype engine.

Bizzarrini designed a dry-sump, 3,465cc, 60 degree V12 engine, with a 77mm bore and a 62mm stroke, boasting twin overhead camshafts per bank, and no fewer than six 38mm Weber down-draught racing carburettors. This prototype V12 was put on a dynamometer on 15 May 1963 and it produced 358bhp at 9,800rpm and 240lb-ft torque at 6,000rpm. Bizzarrini was delighted.

But Ferruccio Lamborghini was appalled. Genuinely appalled. By all accounts Ferruccio was a thoroughly decent man who always settled his debts, but on this occasion he refused to pay Bizzarrini, saying that this prototype V12 engine was not fit for purpose. Ferruccio elaborated that he had specifically told Bizzarrini that this engine was to be used in an expensive, exclusive, and luxurious Gran Turismo road car, and that Bizzarrini's quasi-racing V12 engine's power and torque delivery was far too 'peaky', with too narrow a power and torque band to be employed in a civilised Gran Turismo car.

Having fallen out with Giotto Bizzarrini, Ferruccio saddled Giampaolo Dallara and Paolo Stanzani with the monumental task of modifying Bizzarrini's quad-cam 60 degree V12 engine to his exacting requirements.

First of all Dallara and Stanzani jettisoned the complex and costly dry-sump lubrication system – it would be another thirty-eight years before a production Lamborghini flagship would feature this technology

Gandini's extreme wedge-shaped design is well illustrated here with a single uninterrupted line from the front bumper to the roof.

again – as seen in the 2001 Murcielago 6.2 VT. Standard side-draught Weber carburettors now replaced the original racing carburettors, and the valve timing, valve stems, tappets and distributors were all altered or replaced. Dallara's and Stanzani's modified Bizzarrini engine produced 270bhp at 6,500rpm and 240lb-ft at 4,000rpm, with power and torque curves that were much more to Ferruccio's liking. This was the engine that was fitted to the very first Lamborghini car – the 1963 350 GTV – Ferruccio's long-sought after 'perfect Gran Turismo road car'.

This quad-cam 60 degree V12 remained in production from 1963 until the Murcielago's demise in 2010, during which time it was stretched in just about every dimension, saw fuel injection replacing its original carburettors, revisited dry-sump lubrication, and was subjected to a thousand different modifications. But in its essence it remained the same engine that Bizzarrini had designed. This stalwart V12 engine not only lay at the heart of every pre-Aventador Lamborghini flagship car, but it was also undoubtedly early Sant'Agata's eternal and never-changing soul.

MARCELLO GANDINI AND THE COUNTACH'S REVOLUTIONARY AESTHETICS

The Lamborghini Countach's aesthetics were truly revolutionary for a production road car. In the early 1970s the sight of an angular Countach driving down Chesterfield Gardens in Mayfair would have been akin to a flying saucer landing in Times Square in 2025. And the Countach's aesthetics were genuinely influential – the wedge silhouette can still be seen in today's flagship Lamborghini road car, the 2023 Revuelto, in Lamborghini's Essenza SCV12 track-only hypercar (2020), and in Sant'Agata's Sian FKP 37 supercapacitor-bearing hybrid supercar (2019). Gandini's extreme wedge-shape design heavily influenced other

Above: Gandini's wedge-shaped design demanded retractable headlights and closed-in indicator pods.

Opposite: The Countach's Kammback tail was carried through to the Diablo and the Murcielago.

supercar manufacturers too. Ferrari's doorstop-like Testarossa (1984) was a marked departure from its curvaceous predecessor, while BMW's angular BMW M1 (1978) and Aston Martin's origami-sharp Bulldog (1979) further demonstrated that Maranello, Munich and Newport Pagnell all felt obliged to bow down low before Marcello Gandini's masterpiece. Needless to say, the Lamborghini Diablo and Murcielago body forms were evolutionary derivatives of the Countach's body design.

Ruote O.Z.

Marcello Gandini was working at Carrozzeria Bertone in Grugliasco on the outskirts of Turin when he penned the Countach. Nuccio Bertone and Ferruccio Lamborghini trusted Gandini implicitly and gave him free rein in designing Lamborghini's latest flagship car. Gandini had already won widespread acclaim for designing the Lamborghini Miura (1966), although this success had been tinged by some slight bitterness when Giorgetto Giugiaro, the then Head of Design at Bertone, claimed partial authorship of the Miura. But Gandini had taken over from Giugiaro years before the Countach was even conceived of, so it can be stated without caveat that the Lamborghini Countach's body form was entirely and solely Marcello Gandini's work.

The Countach's spectacular bodywork was revolutionary for a production road car, but it followed in the footsteps of two other Bertone concept cars – the Alfa Romeo Carabo and the Lancia Stratos HF Zero – both also designed entirely by Marcello Gandini. The Carabo debuted at the 1968 Paris Motor Show and had an extreme wedge shape, true guillotine doors, a truncated Kammback tail, a window within each side window, and large air intakes above the rear wheels. Every one of these key Carabo design features would be seen three years later in the LP 500 Countach concept car. The Lancia Stratos HF Zero concept car was even more extreme than the Carabo. First shown at the 1970 Turin Motor Show, this outrageous vehicle had originally been called the Stratoslimite – which translates into 'only limited by the stratosphere'. The HF Zero was even more angular than the Carabo, and with a maximum height of 846mm (33.3in) it really looked like a mobile door wedge. The Countach inherited the HF Zero's angularity, its cliff-like Kammback tail, its uprising and slashed rear wheel arches, and its gaping flank air-intakes, but not its impossibly impractical entry route into the cockpit – which was via a front-hinged flip-up windscreen!

Above: The Countach's red tail-light surround looks like a fighter jet's exhaust in full afterburner mode.

Opposite: Gandini's ultra-sleek original design had to be desecrated with elephant-ear RAM ducts because the Countach's V12 engine needed much more cooling than originally anticipated.

Gandini designed the Countach to be very low, very wide, and relatively short. The Countach LP 500 prototype was particularly striking in side profile – from its sharp nose so very close to the ground, a straight, uprising, diagonal line could be traced, without any interruption for the windscreen, all the way to the roof. From there another completely straight, this time horizontal line demarcated the full length of the roof, after which another straight, slightly down-sloping line ran to the rear of the Countach. At the very back of the car was the final straight line – an abrupt, cliff-like, vertical cut-off that formed this concept car's Kammback tail.

The origami-style concept LP 500 also boasted another design motif. Gandini chose the trapezoid – a flat surface with four straight sides, two of which on opposing sides are parallel – to further emphasise the Countach's futuristic design language. The Countach's front bonnet, windscreen, roof, door glass, quarter glass, engine compartment bonnet, and rear light enclosures (which looked like a fighter jet's red-hot afterburners at full thrust) – all adopted this trapezoidal configuration.

Marcello Gandini also designed the Lamborghini Diablo. And Luc Donckerwolke was honest and brave enough to explicitly state that he took inspiration and design cues from Gandini's Countach and Diablo when he was sketching out the Murcielago's body form on his drawing board.

Both the Diablo and the Murcielago inherited the Countach's party piece – its dramatic guillotine doors. The Countach was the first true production car to feature vertically opening doors, and the name Lamborghini has for decades been synonymous with scissor doors. Today this no longer holds true – Sant'Agata's V10-engined junior supercars have conventional doors, and the Aventador and the Revuelto do not have true guillotine doors – their doors swing outwards as they rise up.

It remains entirely fitting that the revolutionary yet simultaneously aristocratic Lamborghini Countach was the first car to bring this new guillotine door design to the attention of the masses. When Joseph-Ignace Guillotin, the physician and French revolutionary, advocated the guillotine as the most humane and efficient way of executing aristocrats during the 1789 French Revolution, he also unwittingly provided

Above: The Countach's NACA ducts were a much more welcome late addition than the RAM ducts, and would immediately become one of the Countach's most iconic design features.

Opposite: The Countach's revolutionary guillotine doors were carried forwards to the Diablo and the Murcielago, and quickly became synonymous with Lamborghini.

Giampaolo Dallara engineered the Miura and then returned to Sant' Agata to fettle the later Countach variants.

free entertainment for the bloodthirsty public. (At that time common criminals were hanged, but the aristos were spoilt enough to demand special treatment in the form of beheading even as they prepared to meet their Maker.) Beheading by hand and axe was a slow and inefficient procedure, and often prone to accident and mishap. So bloodthirsty revolutionaries and condemned aristocrats alike favoured the guillotine, and with time, and in the name of 'égalité', this highly dubious favour was extended to commoners too. Today, the closest possible 2025 reenactment of Madame Guillotine at her bloodiest-best is a rapidly descending Countach guillotine door (courtesy of a failing gas strut) threatening the gilded neck of its billionaire owner. And the watching masses in Monte Carlo's Casino Square will be no less amused or entertained by this twenty-first century spectacle than their eighteenth-century predecessors were.

UMBERTO MARCHESI'S INTRICATE SPACEFRAME CHASSIS AND THE TIME-TESTED SUPERLEGGERA CONSTRUCTION METHODOLOGY

The Countach's inner skeleton was constructed from an intricate meshwork of hand-welded round-section steel tubes to form a spaceframe chassis, while its outer bodywork was made up of hand-formed and hand-beaten aluminium alloy panels. These two component parts were married together using the Superleggera construction technique.

In a modest industrial unit just off Modena's Via dei Tipografi lies a veritable wizard's den – Marchesi & C. In 1970 Ferruccio Lamborghini commissioned Umberto Marchesi to fashion a spaceframe chassis for his forthcoming flagship supercar. This spaceframe would form the skeleton upon which the Countach's hand-beaten aluminium panels would hang. Marchesi's hand-welded spaceframe chassis, although largely hidden from view, is just as emblematic of the Countach as the Countach's unique drivetrain layout, or its angular body form, or its guillotine doors.

A spaceframe chassis carries many advantages and relatively few disadvantages, and it was therefore unsurprising that Sant'Agata chose to stick with this type of inner skeleton and separate outer bodywork system for the Diablo and Murcielago as well – albeit with evolutionary modifications and ever more composite reinforcements. It was only with the advent of the Aventador with its carbon fibre chassis that Lamborghini finally abandoned the spaceframe chassis for its mid-engined V12 flagships.

A spaceframe chassis is designed such that each chassis tube is only ever subjected to an extension force or a compression force, and never to a bending force. This arrangement provides exceptional rigidity per unit weight. Chassis rigidity is crucial for handling and roadholding, while low weight is important for acceleration, braking and cornering. Torque to weight and power to weight ratios determine just how quickly a car gets off the line and how quickly the car accelerates thereafter. In the 1970s two numbers dominated every supercar sales brochure and every supercar magazine article. Top speed and the time required to accelerate from 0 to 60 miles per hour were venerated statistics that every self-respecting schoolboy could recite out aloud from memory for every Ferrari, Lamborghini, Aston Martin, and Porsche road car. Employing a lightweight but rigid chassis gave Paolo Stanzani his best chance of installing the Countach at the very top of the supercar performance tree.

Every production Countach had a hand-welded, round-tube, spaceframe chassis, while every production Diablo and Murcielago had a hand-welded, square-tube spaceframe chassis. It is of note that the 1971 Countach LP 500 concept car had a Marchesi fabricated

A silver-coloured Dallara-engineered Lamborghini Miura in the foreground with a blue Stanzani-engineered Bugatti EB110 in the background. Stanzani succeeded Dallara as Lamborghini's Technical Director in 1968. Gandini designed both the Miura and the EB110.

semi-monocoque chassis, while the Countach Evoluzione mobile test bed car had a carbon fibre monocoque chassis – but these two one-off cars were the exceptions that highlighted the rule.

While the steel tubes used were relatively inexpensive, welding these tubes together to make the intricate and precisely contoured lattice was a hugely expensive process – expensive in terms of expertise, time and liras. In the early days of Countach production, when demand hugely outstripped supply, Sant'Agata was unable to push more cars out of the factory gates principally because Marchesi & C couldn't find enough expert and experienced welders to make more Countach spaceframes.

The Countach's spaceframe chassis was fabricated from 15mm, 25mm, 30mm, and 40mm round-section steel tubes, all with a 1mm wall thickness. Each Countach spaceframe weighed just 90kg, a number that belies its importance as one of the key components in a supercar which, depending upon model variant, weighed in at between 1,300kg and 1,490kg.

Overlying the Countach's spaceframe chassis, and giving form to this veritable birdcage of naked steel tubes, was an assortment of hand-beaten aluminium body panels. Most of the Countach's body was fashioned from 1.0mm to 1.5mm thick aircraft grade aluminium alloy panels – the notable exception here being the Countach's steel roof. True artisan panel beaters, no less expert than the talented welders at Marchesi & C who hand-fabricated the spaceframe chassis, hand turned and hand beat flat aluminium alloy sheets into voluptuous front wings and rear flanks, and origami-straight bonnets, engine lids, and Kammback tails. During the Countach's production run, Lamborghini's panel beaters were considered to be amongst the world's best – period.

The Countach's aluminium alloy body panels were attached to its spaceframe chassis using the Superleggera construction technique. The word Superleggera translates into 'super-light', and this build method was conceived and pioneered by an early aviator and racing pilot by the name of Charles Terres Weymann. The half-Haitian and half-American Weymann got his inspiration from the way that airframes were manufactured in the earliest days of the twentieth century and brought this technology of a lightweight frame with a fabric covering to the automotive sector. Felice Bianchi Anderloni's and Gaetano Ponzoni's Milan-based Carrozzeria Touring further developed and evolved Weymann's original idea by replacing the fabric with alloy panels, and in 1936 took out a patent on this Superleggera method of coachbuilding. With the Countach, hand-formed aluminium alloy panels were attached to the steel Marchesi spaceframe. The Diablo and Murcielago were constructed in a similar manner, but with carbon fibre and other composite panels increasingly replacing the alloy body panels.

AERODYNAMICS DRIVES THE HOLY TRINITY'S EVOLUTION

The Countach's perennial Achilles heel was its aerodynamics. Despite its futuristic appearance and arrow-like side profile, the Countach suffered from truly dire aerodynamics, and Sant'Agata's engineers had to constantly beef-up each subsequent variant's torque and power outputs simply to compensate for this singular weakness. Especially as almost every bodywork modification and alteration that saw the ultra-sleek LP 500 concept car evolve, through six model variants, into the meretricious Countach 25th Anniversary involved tacking on drag-inducing slats, vents, dams, spoilers and wings – each of which further eroded the Countach's already appalling aerodynamics. This technical crisis was further exacerbated by a sales and marketing challenge. Every new rival that made its appearance during the Countach's unbelievably

Valentino Balboni, Lamborghini's ever-green test driver was the legendary Bob Wallace's protégé. Balboni was involved in developing the Countach, Diablo and Murcielago.

long seventeen-year reign arrived with a wind tunnel-honed body, and their manufacturers trumpeted this advance while specifically denigrating the Countach's dire aerodynamics. The Countach's death knell was finally struck, and the Diablo brought into production, when Sant'Agata belatedly recognised that nothing less than a wholly different body was needed if Lamborghini was to retain its supercar crown. In Darwinian terms, the Countach's aerodynamics rendered it so unfit for purpose that by 1990 the car no longer deserved to survive.

Sant'Agata never ever put a full-scale production Countach through proper wind tunnel analysis and development. Stanzani and Gandini did put a wooden LP 500 scale model into Politecnico di Torino's wind tunnel, and Stanzani, Bob Wallace, and Massimo Parenti did do some real-time, but very rudimentary, aerodynamic analysis of the full-size LP 500 concept car using wool tufts and high-speed photography, but this was all the aero development that the Countach ever received.

The Countach is widely reported to have an aerodynamic drag coefficient (Cd) of 0.42. This was a dreadful figure, even by the standards of the day. The 1966 Lotus Europa had a Cd of 0.29, the 1969 Alfa Romeo Giulia had a Cd of 0.34, and the 1974 Ferrari 365 GT/4 Berlinetta Boxer had a Cd of 0.35. But even this Countach Cd figure of 0.42 was wildly optimistic – it was derived from an ultra-sleek and hyper-smooth wooden scale model of the LP 500 concept car. Every production Countach on the other hand was further hamstrung by some or other combination of drag-inducing and aerodynamically inefficient vents, ducts, air-dams, front spoilers, rear wings, pentagraph windscreen wipers, and door-mounted mirrors.

Leaving aside the awfulness of the Countach's supposed 0.42 Cd for a moment, the Countach had an even more pressing aerodynamic problem. In real-life driving, what actually matters is not the Cd, but the CdA. A car's CdA, which is calculated by multiplying its aerodynamic

Horacio Pagani oversaw the design of the Countach Evoluzione and the aesthetics of the 25th Anniversary Countach.

Maurizio Reggiani was Automobili Lamborghini's Chief Technical Officer for fifteen years. He worked on the Canto and oversaw the development of the Murcielago.

drag coefficient (Cd) with its frontal area (A), is a measure of the total aerodynamic drag that a car is burdened with, and determines the car's acceleration, top speed, fuel consumption, and noxious emissions. The CdA also influences a car's cornering performance, its braking potential, and whether or not it meets the existing environmental protection standards required for homologation. And the Countach, with its wide bonnet, huge windscreen, and massive Pirelli P7 and P Zero rear tyres (from the LP 400 S onwards – the earlier cars employed much narrower Michelin tyres) boasted an enormous frontal area and so had a truly ghastly CdA.

All this is not to say that Stanzani and Gandini ignored the Countach's aerodynamics. On the contrary, they were acutely aware that the design brief that Ferruccio Lamborghini and Nuccio Bertone had handed down to them demanded world-beating acceleration and top speed. Specifically, a maximum speed of at least 300km/h twinned with exemplary high-speed stability – so not a simple ask. The Lamborghini Miura's high-speed, front-end lift and snap oversteer were things that Ferruccio desperately wanted to see the back of. These two flaws, while great for Lamborghini's accident and repair department's balance sheet, gave Ferruccio's young supercar company a much unwanted reputation for unstable, difficult-to-control, and dangerous cars. It was therefore a key priority that the Miura's replacement should be aerodynamically stable. While reducing drag was vitally important to Stanzani and Gandini, their overarching ambition was to reduce lift. And when Bob Wallace first took the LP 500 prototype car out for high-speed testing, he returned back to the factory delighted – the car had not shown any signs of front or rear end lift. Mission accomplished – at least with respect to that particular all-important design brief.

Stanzani and Gandini's second design brief – which concerned top speed and acceleration – called for reducing aerodynamic drag as much as possible. In real-life driving aerodynamic drag is the chief force that works against the power and torque that a car's engine develops. Aerodynamic drag completely eclipses tyre-road surface frictional forces, tyre deformation forces, and ground deformation forces – especially at high speeds.

Crucially the relationship between aerodynamic drag and speed is non-linear. A car's aerodynamic drag is in fact directly proportional to the square of its speed. Furthermore, the power that a car's engine has to produce to overcome aerodynamic drag alone (i.e. discounting all the other frictional forces mentioned above) is proportional to the cube of the car's speed. This means that for a Countach to double its speed from any given baseline, its engine has to produce eight times (the cube of two) as much power.

With a non-negotiable design brief for a minimum top speed of 300km/h, Stanzani and Gandini now had to tackle their second priority of minimising aerodynamic drag. They did this by employing shape optimisation techniques – ensuring that the Countach's sharp nose was the first thing to encounter the on-coming airstream, thereby minimising aerodynamic turbulence and drag. Going against first impressions, the shape of the front of a car has far less influence on its aerodynamic efficiency than the shape of its rear. This is because the shape of the rear determines how the boundary layer of the airstream detaches itself, which in turn determines the resultant wake, and so the overall drag to which the car is subjected. To this end Gandini and Stanzani chose a Kammback tail for the Countach. A Kammback tail's abrupt vertical rear cut-off generates an area of turbulence immediately behind the car, which in turn reduces aerodynamic drag.

Stanzani and Gandini tried to further minimise the LP 500 Countach's Cd and CdA by employing a partially flat undertray, and by deliberately omitting drag-inducing windscreen wipers, external mirrors, and radio antennae. Legislation, however, demanded that the first two items be

Ferruccio Lamborghini looks down on a Diablo at the eponymously named Museo in Casette di Funo, Italy. The rear end of a 350 GTV replica can be seen to the right of the Diablo.

Doors aloft – the Countach and Murcielago sandwich Lamborghini's Holy Trinity.

included on every production Countach, while Lamborghini replaced the fixed aerial found in early production cars with a retractable aerial at the earliest possible opportunity – simply to minimise aerodynamic drag as much as possible. Despite this degree of attention to detail, and despite all the time and effort Gandini and Stanzani put into honing the Countach's body form, the Countach's aerodynamics remained its perennial Achilles heel, and was what finally rang the death knell for the Holy Trinity's progenitor car.

CHAPTER 3

THE DEVELOPMENT OF THE COUNTACH AND THE EVOLUTION OF ITS MODEL VARIANTS

In the period between the debut of the Countach LP 500 prototype in 1971 and the last Countach rolling off Sant'Agata's production line in 1990, the Countach bloodline spawned six production variants and one particularly notable mobile test platform variant. The six production variants were the LP 400, LP 400S, LP 500S, 5000 Quattrovalvole, 88 ½ Quattrovalvole, and the 25th Anniversary. The need to evolve each of these production variants into the next was forced upon Sant'Agata by a combination of customer expectation, new technology becoming available, inter-manufacturer rivalry, government legislation, road safety regulations, fuel availability, and fuel efficiency issues brought on by wars and revolutions, and environmental considerations, particularly air pollution. The Countach Evoluzione mobile test bed was a one-off experimental car, the findings from which would go on to inform the last of the Countach variants as well as the Diablo and the Murcielago.

THE PROTOTYPE COUNTACH LP 500

The bright yellow, ultra-sleek, and origami-sharp Countach LP 500 prototype was unveiled on the Bertone stand at the 41e Salon Auto Geneve on 11 March 1971. It was presented as a concept car that might or might not make it into production. Sant'Agata's internal designation for this revolutionary car was Project 112, and it had two prime design objectives. Firstly, Project 112 was conceived, designed, and built to banish into the history books the high-speed, front-end lift that had plagued Lamborghini's last flagship car, the Miura. Secondly, Project 112 had to be capable of 300km/h. At the Geneva Motor Show, Lamborghini loudly and triumphantly claimed that the LP 500 could meet this latter performance objective, although this was never put to the test.

Ultimately, the LP 500 prototype, although very promising in so many respects, was nowhere near ready to go into series production. It would

The unique Countach LP 500 prototype parked by the factory gates.

need another three further years of extensive modifications, intensive alterations, and wholesale replacement of key parts before a Countach that was fit for purpose could be put into routine production and sold to the many well-heeled and exacting customers patiently but excitedly awaiting delivery. Although Ferruccio refused to confirm that his latest flagship prototype would make it into series production, countless blank cheques were thrust into the clenched fists of the Lamborghini personnel manning the Bertone stand that March day – such was the allure of Sant'Agata's latest jaw-dropping offering.

There was good reason for Ferruccio Lamborghini's reticence – he knew perfectly well that even the very fundamentals of the LP 500 – its engine and its chassis – had to be completely changed, and its body form extensively altered, before this prototype could be presented as a road car.

The LP 500 prototype took its name from the orientation and the displacement of its Bizzarrini V12 engine. The LP part of its designation stood for Longitudinale Posteriore, while the 500 part of its name indicated its Cinque Litri cubic capacity. Paolo Stanzani's original specification for the revolutionary Countach called for a 4,971cc engine with a bore of 85mm and a stroke of 73mm, fed through six side-draught Weber twin-choke carburettors, boasting a high compression ratio of 10.5:1, a peak power of 440bhp at 7,400rpm, and a peak torque of 366lb-ft at 5,000rpm. But only one such engine was ever built, and since this unique engine was not ready in time for the Geneva Show, the LP 500 prototype shown there was powered by a much more prosaic 3,929cc Bizzarrini V12 engine. When this new 4,971cc engine did finally get built and installed, Stanzani and Gandini were devastated to discover that it put out a whole lot more heat than they had originally anticipated, which in turn caused major engine bay and drivetrain cooling problems, as detailed below. Bob Wallace affectionately called this one-off 4,971cc engine 'the big-bore, end-of-the-world, prototype large engine', but his enthusiasm for it did not survive after it tried to kill him. Wallace regularly tested the LP 500 at high speed on the autostrada, and on sections of the old Mille Miglia route including the Futa and the Raticosa passes, but on one test drive this sui generis engine self-combusted and exploded into a thousand pieces without warning. The resulting fire not only threatened Wallace's very existence but also damaged a substantial part of the prototype's chassis, bodyshell, and interior. The LP 500 was salvaged and repaired, but it and future Countach test cars, as well as the early production variants – the LP 400 and the LP 400S – all reverted to using the time-tested and proven 3,929cc Bizzarrini V12 engine.

Another of the Countach's key components was similarly changed out prior to the car being put into series production. The LP 500 prototype was presented at Geneva with what Lamborghini called a monocoque chassis, but which might more accurately have been called a semi-monocoque/partial spaceframe chassis. This too was quickly jettisoned for the beautiful and intricate, Marchesi-fabricated, round-section, hand-welded, tubular spaceframe chassis that instantly became one of the defining features of the Countach.

The LP 500 prototype's stunning body had to sacrificed too – in the name of engine bay and drivetrain cooling. Even with the smaller 3,929cc engine, let alone the one-off 4,971cc engine, there was insufficient ambient temperature airflow to remove the huge quantities of heat generated by the V12 engine. Sant'Agata had encountered this very same problem with the Miura previously, and Stanzani and Gandini knew from bitter experience that only major surgery would resolve this important and pressing issue.

So major surgery it was. To keep the LP 500's aerodynamic drag coefficient as low as possible Stanzani had orientated the prototype's

A 1974-vintage Countach LP 400, otherwise known as the Countach Periscopio.

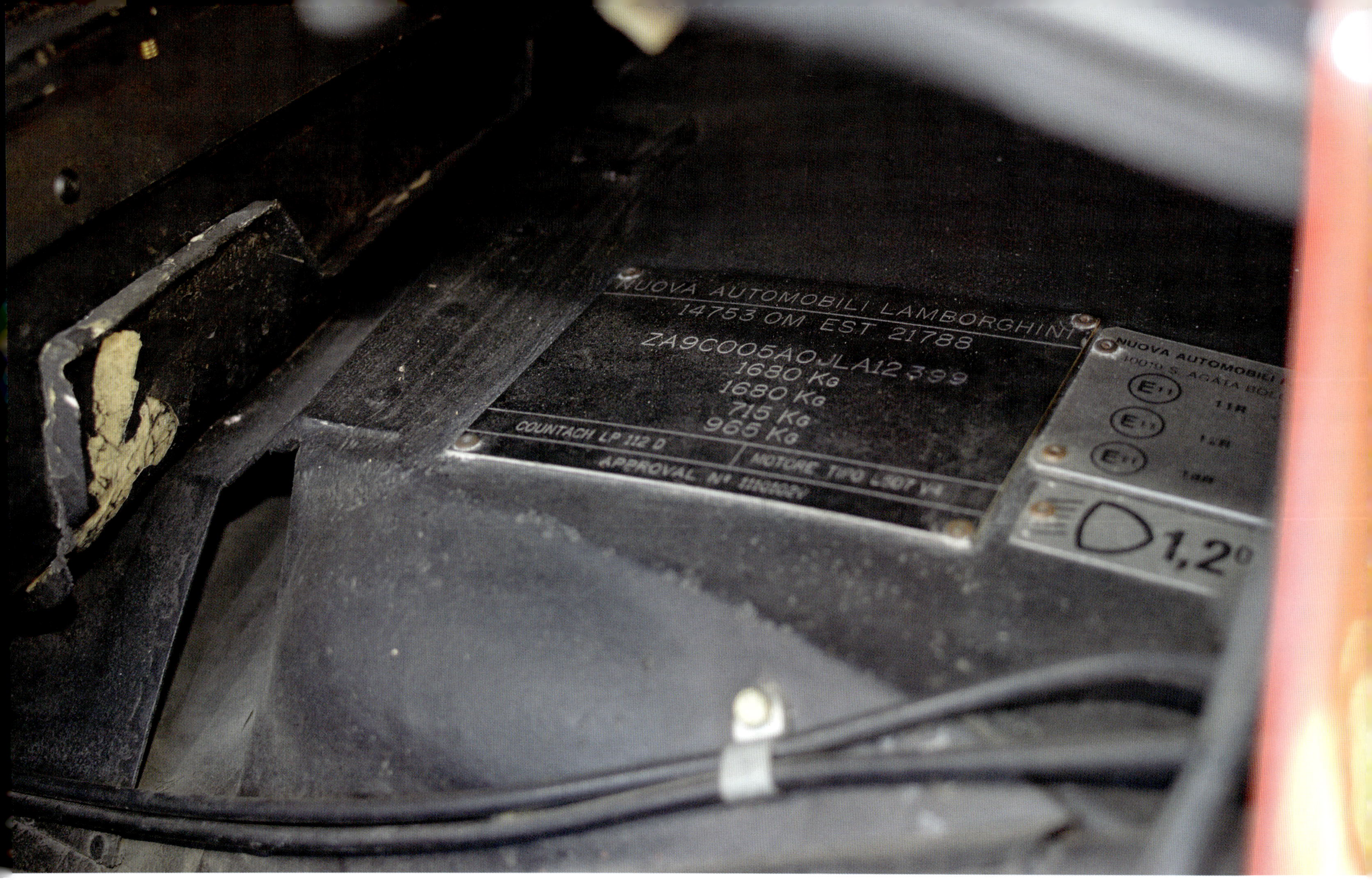

A Countach's Vehicle Identification Plate.

radiators longitudinally and sideways, and Gandini had designed sleek and elegant vents that resembled sharks' gill slits to clothe these radiators. This beautiful arrangement now had to be abandoned. The radiators were turned through 90 degrees so that they now faced the oncoming airstream directly, and they were now housed within gargantuan RAM airducts that were better able to funnel outside air into these radiators. These RAM ducts were affectionately nicknamed Elephant Ear ducts for obvious reasons, but they did nothing for the Countach's aesthetics. Beauty and aerodynamic efficiency had been forced to bow before the over-arching need for adequate and efficient heat dissipation.

If the RAM ducts were a touch graceless, the NACA ducts that Stanzani and Gandini were compelled to carve into the pre-production test cars' doors and rear flanks were glamorous and refined additions. The term NACA duct derives from the institution – the National Advisory Committee for Aeronautics, the forerunner to NASA – which first conceived and developed this type of low-drag air inlet. Scientists there found that a progressively widening and deepening submerged air inlet, with shallow sides and a ramp angle of less than 10 degrees, allows for large volume air inflow with minimal induced aerodynamic drag. The NACA ducts in the Countach extended backwards from the rear quarter of the car's guillotine doors to the front quarter of the car's rear flank. These futuristic-looking NACA ducts not only increased the airflow into the engine compartment by 30%, but also housed the push buttons for opening the guillotine doors. They too instantly became one of the Countach's signature features.

The LP 500 prototype's Gianni Rogliatti-inspired, spacecraft-like interior was another feature that did not survive past the concept car stage. The single-spoke steering wheel, the minimalist dashboard, the hammock seats with chocolate cube pads, the digital speedometer and tachometer, the Donnelly Corporation periscope-style rear-view mirror – all had to be either modified or abandoned as Project 112 evolved along the path to series production.

At the start of Project 112 Ferruccio Lamborghini, Paolo Stanzani, and Marcello Gandini were all in complete agreement that the Countach should be a no-expense spared, clean sheet design, which exploited every opportunity for putting blue-sky thinking into practice.

For Stanzani, a long-standing admirer of Lotus's Colin Chapman and a keen adherent of his adage 'Adding power makes you faster on the straights, Subtracting weight makes you faster everywhere', using very lightweight, but very costly materials in the Countach's construction was a necessary evil. To this end, Stanzani used Elektron, a hugely expensive but very light magnesium alloy, for the LP 500's engine sump, engine mounts, oil filter housing, brake callipers, and suspension uprights. Sadly, during the course of the LP 500 prototype's test programme it was discovered that the relatively fragile Elektron could not withstand some of the forces imposed upon it. As a direct consequence of this discovery later pre-production cars and all production Countachs had their engine sumps, for instance, fabricated from aluminium alloy instead of Elektron.

As Wallace, Stanzani, and Parenti road-tested the LP 500, a whole host of lesser problems surfaced, leading to further modifications and alterations being carried out sequentially with each new revelation. But in essence this trio felt that they had a real winner on their hands in the LP 500 Countach. All the fundamentals were correct, and every major and minor issue could be sorted out – even if this demanded painstaking, time-consuming, and expensive work. Their faith in this Countach prototype car and its underpinnings was validated when Wallace and Stanzani drove the LP 500 down to Sicily to watch the Targa Floria. This gruelling, long-distance trip involved tackling some of Italy's toughest and most testing road surfaces.

This Countach LP 400 S boasts flared wheel arches and Bravo-style wheels.

On their return from spectating at this Sicilian road race they reported in detail to Ferruccio Lamborghini, who then made the momentous decision to put the Countach into series production. This singular decision would keep Automobili Lamborghini S.p.A. afloat and alive for the next seventeen years.

Two pre-production Countachs were built following Ferruccio's brave decision, and much confusion has issued forth because of the colours of these first three Countach cars. The LP 500 prototype was yellow, and the first pre-production Countach was initially painted red but then repainted green. The second pre-production car was yellow. So three different cars and three different colours, but one car wore two different colours and two cars shared the same colour. Cue confusion.

The unique and priceless LP 500 prototype came to a very sad but necessary end when, in order to gain Type Approval, it was barrier crash-tested against a concrete block at the Motor Industry Research Association's test facility in the United Kingdom on 21 March 1974. This crash test proved just how strong the LP 500 was – its windscreen remained intact, its cockpit area was left largely unscathed, and its guillotine doors did not buckle and could be opened as normal. With the successful completion of this test the Countach earned European Type Approval and an E-mark and series production could now begin.

The Countach LP 400

On 11 April 1974 the first production Lamborghini Countach passed through the factory gates destined for an incredibly lucky Milanese customer. This was barely two years after the 1972 Bolivian Economic Crisis, just months after the 1973 Yom Kippur War, and halfway through the consequent OPEC-fermented oil crisis. The fallout from Bolivia's mismanaged economy would ripple across the Atlantic and the Mediterranean and eventually force Ferruccio Lamborghini to sell his eponymous supercar company. Meanwhile, the highly varied and largely negative repercussions from this particular Middle Eastern war would reverberate through the global automotive industry for decades to come. Launching a 196mph supercar at the same time as the United States Congress was passing legislation for a blanket speed limit of 55mph verged on the ridiculous, especially when taken in the context of fuel rationing and sky-high petrol prices. Especially since there was, and remains, no known way of defying the fundamental laws of physics that decree that the aerodynamic drag on a vehicle is proportional to the square of its speed, and that the power (and therefore the fuel combusted) needed to overcome this drag is proportional to the cube of the vehicle's speed.

But mere logic and pragmatism could not tame the lust for a Countach, and Lamborghini very quickly found itself with a twelve-month long waiting list for its latest flagship supercar. Supply rather than demand was always the issue with the Countach LP 400, and there were simple but not easily resolved reasons why Sant'Agata could not just turn on the taps. Firstly, Marchesi & C could not find enough expert welders to fabricate the Countach's intricate spaceframe chassis. Secondly, artisans capable of hand-beating flat aluminium alloy sheets into the Countach's angular but simultaneously voluptuous body panels were a rare and not easily found breed. Ditto the engine builders, leather workers, and production line assemblymen – all these people were highly skilled craftsmen and women. It also took time and effort to train the people who operated the highly specialised machinery needed to build a Countach, especially as many of the tasks involved working to exceptionally fine tolerances. A further bottleneck in the supply of these earliest Countachs came in the form of the disruptive industrial unrest and union activity that was rocking the whole of Italy during the 1970s. Only about 150 Countach LP 400 cars were produced during

A Countach LP 400 S with raised retractable headlight pods and driver's side guillotine door.

This yellow Countach LP 400 S is gifted additional presence by the sheer width of its delta-shaped rear wing and its vast Pirelli P7 tyres.

The Full Monty – a black Countach LP 5000 S (earlier versions were called LP 500 S) with flared arches, gold-coloured wheels and that unmistakable drag-inducing, speed-sapping, rear wing.

this variant's four-year lifespan, which ran from 1974 to 1978, making the LP 400 the second rarest production Countach by some margin (Only 145 88 ½ QV cars were produced, albeit over a much shorter timeframe).

The LP 400 remains the Countach that most closely mirrors Stanzani's and Gandini's original LP 500 prototype. It employed the stalwart normally aspirated, two valves per cylinder, quad camshaft, wet sump, 3,929cc, Bizzarrini V12 engine, which had a stroke of 62mm and a bore of 82mm. This 245kg aluminium alloy engine had hemispherical combustion chambers, cast-iron cylinder liners, lightweight forged alloy pistons, and forged steel connecting rods. It ran a compression ratio of 10.5:1 to produce a peak power of 375bhp at 8,000rpm, and a peak torque of 266lb-ft at 5,500rpm. Two 60-litre fuel tanks hidden within the car's side sills fed two electric Bendix fuel pumps, which in turn fed six side-draught, twin-choke, Weber 45DCOE carburettors. Each fuel tank had its own fuel filler inlet hidden within the corresponding side's NACA duct and a narrow bore pipe connected these two tanks. A Countach could therefore legitimately take up the whole lane of any petrol station forecourt as fuel could be pumped into the right- and left-side fuel tanks simultaneously.

The LP 400 had a dog-leg, five-speed, manual gearbox, a ZF limited-slip differential, and used the same hydraulically actuated, single, dry-plate, Fichtel & Sachs clutch system found in the 1970 and 1971 Le Mans-winning Porsche 917 endurance racers. The engine bay and drivetrain cooling problems that had plagued the LP 500 prototype were solved in the LP 400 by employing twin pressurised radiators with a cross-over connection as well as electric radiator fans activated by a thermostatic switch.

Aping, and for very good reason, contemporary Formula 1 racing cars, the LP 400 used classic, parallel, unequal-length double wishbones with co-axial dampers and springs (twin on each side at the rear because of the heavy V12 engine), and a front anti-roll bar. The rack and pinion steering was unassisted, but the dual-circuit Girling braking system was hydraulically operated and vacuum servo-assisted, and worked on 267mm ventilated front and rear discs. Cast magnesium alloy Campagnolo wheels were wrapped in 205/70 VR 14 front and 215/70 VR 14 rear Michelin WXW tyres. Stanzani had originally conceived the Countach around Pirelli's forthcoming, highly advanced, gargantuan, and ultra-low profile tyre, but sadly the P7 tyre was not available in time for the LP 400's debut.

During the LP 400's short lifespan numerous small modifications and alterations had to be carried out simply to keep the car on the road and to make it liveable with. Early on Sant'Agata discovered that the Elektron magnesium alloy castings were prone to salt-induced corrosion and the factory quickly replaced these with aluminium alloy replacements, which were less corrosion-prone but unfortunately also heavier. Similarly, Lamborghini's engineers had to urgently replace the LP 400's wheel bearings with sturdier, redesigned items when the first cohort of Countach owners reported that the originals could not cope with the stresses and loads imposed by the LP 400's 266lb-ft of torque and 375bhp on a vehicle with a 1,300kg kerb weight.

The LP 400 is commonly referred to as the Countach Periscopio. This is in recognition of its distinctive roof tunnel that Marcello Gandini designed to accommodate the periscope-style rear-view mirror that was meant to be yet another revolutionary feature of this startling car. Sadly, this mirror system (designed by the Donnelly Corporation) could not be made to work in the Countach, and all the production cars therefore had conventional rear-view mirrors instead. The Countach Periscopio's roof tunnel lends it a hint of Zagato's signature 'double-bubble' roof, so giving the LP 400 an added soupcon of exoticism and exclusivity.

Ultimately the 196mph LP 400 with its 0–100km/h acceleration time of 5.6 seconds (claimed performance figures) completely eclipsed its contemporary mid-engined rivals – the Ferrari Boxer, the Maserati Bora, and the De Tomaso Pantera. It took the supercar world by storm courtesy of its technical specification and engineering integrity, its spacecraft-like body, and its spectacular on-road performance. But even all this was not what Paolo Stanzani had originally envisaged for Lamborghini's flagship car. And neither had the forces that drive supercar evolution forwards magically disappeared. Hence there was more, much more, to come.

The Countach LP 400 S

The LP 400 S, which was first presented to the public at the 1978 Geneva Motor Show and which continued in production until 1982, more accurately reflected what Stanzani had originally planned for the Countach. The LP 400 S in fact owed its very existence to a humble tyre, albeit a revolutionary and long-awaited tyre.

Paolo Stanzani left Automobili Lamborghini S.p.A. in 1975, barely a year after the ultra-low profile, tubeless, radial ply, Pirelli Cinturato P7 tyre finally came to market. Way back in early 1971 Stanzani and Ferruccio Lamborghini had come to an agreement with Pirelli that Sant'Agata would be one of the first, if not the first, recipients of the P7 tyre. Stanzani had originally conceived, planned, and developed his avant-garde Countach supercar around this equally state-of-the-art tyre, and he was immensely disappointed and frustrated when Pirelli was unable to produce it in time for the unveiling of the LP 500 prototype, or the debut of the two pre-production Countachs, or the launch the LP 400 variant. Denied the Pirelli P7, Stanzani would also forever be denied the privilege of elevating the Countach to its highest performance potential. This would be left to Stanzani's replacement, Giampaolo Dallara, who would also oversee the extensive redesign of the Countach's bodywork and carry out the requisite engineering revisions needed to accommodate and exploit the very wide P7. Walter Wolf, an outsider, would almost single-handedly finance these modifications and alterations, and in doing so exert a disproportionate influence on Automobili Lamborghini S.p.A.'s future development, and indeed survival.

The P7 was developed by Pirelli's Motorsport Division and featured a novel tyre compound, a unique thread pattern, a 'zero-degree' nylon belt, and special nylon plies in addition to the more usual steel plies. Supercar engineers were excited by the potential performance benefits that the P7 tyre might bring, but supercar designers and supercar owners were excited for more prosaic and base reasons. Contemporary racing cars had been using ever wider and ever more low-profile tyres for some years now, and Lamborghini owners were keen that their expensive supercars emulated these racing cars whenever and wherever possible. For their part, supercar designers were not about to complain about wide, low-profile tyres either, as such tyres made designing dramatic, ground-hugging, UFO-like cars just that much easier. But the very wide P7 came with one major disadvantage – it made the Countach's already awful CdA (aerodynamic drag coefficient multiplied by frontal area) even more dire. But this was seen as a necessary evil and a relatively minor inconvenience. Taken in the round, the P7's debut in 1974 was universally welcomed, especially at Sant'Agata.

The conception and development of the LP 400S was driven forwards by three quite disparate forces. Firstly, Lamborghini's easily bored clientele were now already looking for something new. Secondly, Sant'Agata's engineers were keen to maintain their technological edge over Ferrari, Maserati, and other rivals. And thirdly, Automobili Lamborghini S.p.A's marketing department was desperate to retain Lamborghini's title as the world's most avant-garde supercar manufacturer. The stars were

The mille-feuille astride the Countach 5000 Quattrovalvole engine compartment bonnet is also known as the pillbox.

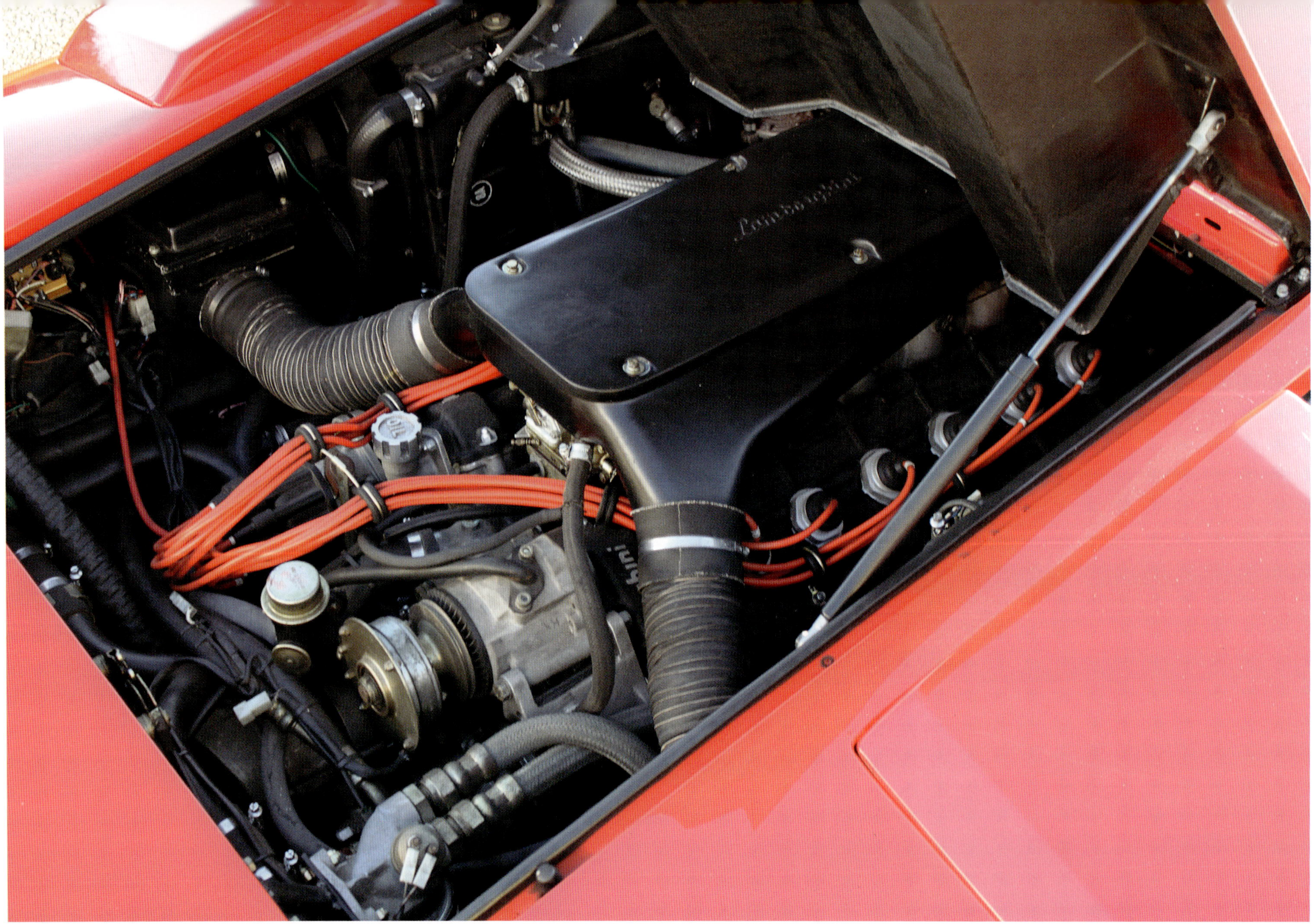

5,167 cc of Sant' Agata's finest Bizzarrini-vintage Quattrovalvole served up in a Countach bodyshell.

therefore aligning perfectly for Sant'Agata to extensively re-engineer and redesign the Countach around the newly available Pirelli P7 tyres.

The only flies in the ointment were Automobili Lamborghini's perilous financial condition, and the fallouts from the Yom Kippur War – namely high petrol prices, limited fuel availability, and low speed limits. By the mid-1970s Ferruccio Lamborghini had sold his stake in the company and Georges-Henri Rossetti and Rene Leimer were disinterested, financially straitened, owners who would not, or could not, invest further in their newly acquired supercar business. And re-engineering and redesigning the Countach for the P7 tyre was a hugely expensive business. Enter stage left the Austrian-born, Formula 1 team-owning, oil and manufacturing tycoon Walter Wolf. A genuine petrolhead and serial Countach owner, Wolf bankrolled Automobili Lamborghini S.p.A. into rehiring Giampaolo Dallara as a consultant to evolve the LP 400 into the Pirelli P7-shod LP 400 S.

Dallara would later describe the P7 tyre as 'The greatest single breakthrough in motoring history', but this meticulous engineer would first carefully evaluate, test, and comprehensively revise the Countach, before bestowing such a fulsome accolade on the P7.

In early 1975 Dallara started re-engineering Countach chassis 1120148 for Wolf. His starting points were the 225/50 VR P7 tyres mounted on 8.5J × 15in wheels for the front axle, and the 345/35 VR P7 tyres mounted on 12J × 15in wheels for the rear axle. To accommodate these new tyres the Countach's whole suspension geometry had to be radically rejigged with new shock absorber pick-up points, revised dampers with different compression and rebound rates, and custom springs with altered tension and compression rates. For the rear suspension, the previously employed reversed A arms were jettisoned in favour of twin parallel arms. The Pirelli P7 tyres endowed the LP 400 S with much higher cornering potential, so stronger wheel hubs were now needed. To accompany these changes a new style of sand-cast magnesium wheel, highly reminiscent of the telephone dial wheels first seen on the 1974 Lamborghini Bravo concept car, was introduced. Larger brake discs (300mm front and 284mm rear), a new steering box, and modified brake callipers were all employed in this latest Countach variant, while the handbrake lever in left-hand drive cars and the accelerator pedal hinge in all cars were relocated.

The bodywork changes that the LP 400 S sported were truly dramatic. The feminine delicacy of the LP 400 was replaced by aggressive masculine aesthetics that would go on to define the Countach for the rest of its production lifetime. Two such signature features would henceforth feature on all Countachs, while the third such signature feature was an optional extra. For road homologation the wide Pirelli P7 tyres had to be covered by the car's bodywork, and Dallara chose to draw attention to the Countach's super-wide, ultra low-profile P7s by tacking on exaggeratedly flared front and rear wheel arches. Dallara also took this redesign opportunity to introduce a prominent front spoiler that wrapped itself around the whole of the lower part of the front of the car, thereby joining up the leading edges of both front wheel arches. The new optional extra was a huge rear wing that would go on to become as much of a Countach signature piece as its guillotine doors already were. The low and sharp new front spoiler married up beautifully with the high and angular new rear wing to give the LP 400 S an even more arrow-like profile than its already impossibly sleek predecessor, but these three additions came at a high price – additional aerodynamic drag, an increased frontal area, and extra weight. The LP 400 S's acceleration and top speed would inevitably be affected, and inescapably be compromised. The rear wing in itself induced sufficient additional drag to reduce the Countach's top speed by around 10mph, and many owners of this and subsequent model variants, your authors included, would choose to abstain from having a rear wing – for aesthetic as well as performance reasons. No lesser a person than Bob Wallace stated without prevarication that the Countach never suffered from rear-end lift and that a Countach rear wing was therefore an unnecessary and performance-sapping ornament.

Relatively minor changes to the LP 400 S throughout its four-year lifespan gave rise to an unofficial classification of this model variant into three subvariants:

- Series 1: Telephone dial wheels and Stewart-Warner dashboard instruments
- Series 2: Concave dished wheels and Jaeger dashboard instruments
- Series 3: Also known as 'High Body cars' with raised suspension and increased cockpit headroom

All LP 400 S cars used the same 3,929cc V12 engine as its predecessor, but Lamborghini was now more candid about this engine's true power and torque outputs. Sant'Agata revised these figures, as well as the LP 400 S's performance figures, downwards to: 353bhp at 7,500rpm, 267lb-ft torque at 5,500rpm, and now only claimed a top speed of 181mph and a 0–100km/h time of 5.9 seconds. But the LP 400 S's adoption of the Pirelli P7 tyre hugely improved its grip, roadholding, and handling, and the LP 400 S was universally praised as a significant improvement over its predecessor.

The Countach LP 500 S

The Countach LP 500 S, which made its debut at the 1982 Geneva Motor Show and continued in production until 1985, was defined by its much-altered, larger-displacement Bizzarrini engine. Dark forces were massing at Sant'Agata's gates crying out for the Countach's supercar crown, and the 3,929cc engine that had given such sterling service in the Countach ever since the LP 500 concept car's 'big-bore' 4,971cc engine had self-combusted during early high-speed prototype testing, was now no longer good enough. Evolutionary forces demanded more power and more torque if the Countach was to remain the fittest of contemporary supercars. Essentially, the Countach's drivetrain had to be extensively modified and upgraded if it was to survive against the onslaught that Lamborghini's rivals – principally Ferrari, Aston Martin and Porsche – were assembling against it in the shape of the BB 512i, the V8 Vantage, and the 911 Turbo.

If the Countach was being threatened, so too was the Lamborghini supercar company itself. Lamborghini, which had suffered significant fallout from the OPEC Oil Crisis of 1973 and the subsequent 1974 stock market crash, had been further affected by the repercussions of the 1978–1979 Iranian Revolution, the 1980 Iran-Iraq war, and the double dip 1980 recession. Automobili Lamborghini had lurched from one financial crisis to the next, but in July 1980 two young, albeit financially very well-endowed, saviours appeared in the form of Partick and Jean-Claude Mimran. Soon thereafter, in May 1981, they bought the company outright and renamed it Automobili Ferruccio Lamborghini S.p.A. With a substantial family fortune behind them, they were able to provide Sant'Agata with an urgently needed initial injection of capital simply to prevent it going to the wall. But the Mimran brothers were not simply profligate spoilt rich kids. They were also astute businessmen and recognised that the Countach was Lamborghini's lifeline and the key to the company's survival, and that they had to urgently remake and improve the model so it remained the fittest of the supercars.

One of the best decisions that the Mimran brothers took was to engage Giulio Alfieri – not only as the company's General Manager but also as its Chief Engineer. Alfieri was one of Italy's finest and most influential

A treasured model of the one-off, now-demised, Countach Evoluzione mobile test platform.

Only fourteen right-hand drive 88 ½ Countach 5000 Quattrovalvole cars were ever produced.

road and race car engineers of the twentieth century – he was responsible for the legendary Maserati 250F, the equally iconic Maserati Tipo 60 and 61 Birdcage racers, and the stunning Maserati Bora, Khamsin, and Ghibli road cars. Alfieri immediately recognised that the Lamborghini 3.929cc engine was time expired, especially in the context of the huge additional aerodynamic drag that the LP 400 S variant's new bodywork bestowed upon it. Nothing less than a comprehensive reworking of the Bizzarrini's V12 engine was now needed – tinkering around the edges would not be sufficient. This redeveloped engine had to produce more power and torque, offer better drivability and reliability, and be less polluting and more fuel efficient. Alfieri considered utilising electronic fuel injection and introducing turbocharging, but quickly dismissed both on the grounds that the former would not liberate sufficient additional power or torque, while exhaust gas-driven forced induction would bring irresolvable problems with heat dissipation, engine cooling, and reliability.

Alfieri chose instead to increase the Bizzarrini V12 engine's displacement to 4,754cc. This would neatly swerve around all the above-mentioned problems while also retaining the much-coveted linear throttle response that is only available from a naturally aspirated engine. Alfieri did not want to sully Lamborghini's long heritage of engineering purity with forced induction, or by replacing the Countach's famous sextet of carburettors with something as inelegant and unromantic as fuel injection. So he instead lengthened the engine's stroke to 69mm, widened its bore to 85.5mm, decreased the compression ratio to 9.2:1, employed a Marelli electronic ignition system, altered the shape of the combustion chambers, and modified the engine's lubrication system.

With these changes Lamborghini now claimed that its V12 engine produced a reliable 375bhp at 7,000rpm and 303lb-ft torque at 4,500rpm. Note that not only were both these outputs higher than what the previous engine had produced, but that they were also liberated lower down in the rev range – an objective vindication that Alfieri had succeeded in making this iteration of Bizzarrini's V12 engine more road driveable and more user friendly than its predecessor. This engine was also about 2.5mpg more fuel efficient, enabling Sant'Agata to dubiously claim the moral high ground against its supercar rivals – with a flagship car boasting a parsimonious 14.5mpg!

Alfieri also increased the LP 500 S's front track by 40mm, but other alterations introduced were trifling in nature: some extra crash padding, changing the shape of the external mirrors, new door card linings, and minor alterations to the cockpit's central console and switches.

As the LP 500 S's production run came to a close it was renamed the Countach 5000 S. Sant'Agata claimed that the LP 500 S/LP 5000 S could reliably reach a top speed of 180mph, and that it could repeatedly and dependably demolish the 0–100km/h dash in just 5.6 seconds. Evolutionary forces had once again triumphed and the supercar crown was still Lamborghini's.

THE COUNTACH 5000 QUATTROVALVOLE

The Countach 5000 Quattrovalvole made its debut in 1985 and continued in production until it was replaced by the 88 ½ Countach in mid-1988.

In 1984 Sant' Agata's nemesis, Maranello, had sought to overturn the supercar hierarchy by launching the Ferrari Testarossa. Ferrari had identified the Countach's Achilles heel with laser-like precision – its aerodynamics – and then targeted this weak point mercilessly. Although Maranello employed Emanuele Nicosia to draw the Testarossa's body form, it ensured that the Testarossa's design team was led by a trained

United States Department of Transport legislation meant that this American-resident 88 ½ Countach 5000 QV had to sport ungainly front and rear aftermarket bumpers.

aerodynamicist, Leonardo Fioravanti. Fioravanti's meticulous detailing of the Testarossa's exterior shape, and especially its numerous slats and vents and protuberances, meant that Maranello's new flagship boasted an aerodynamic drag coefficient of just 0.36 compared to the Countach's Cd of at least, and in all probability a lot more than, 0.42. Sant'Agata now faced a formidable new adversary, and was once again under serious threat.

Giulio Alfieri did not have the financial resources to design and put into production a completely new model to counter Ferrari's Testarossa. Instead he had to again update the already decade-old Countach, and on this occasion he once again resorted to Sant'Agata's time-tested method of keeping up with its rivals – increasing the Bizzarrini V12 engine's power and torque outputs. Alfieri, against his better instincts, was forced to seriously reconsider turbocharging Sant'Agata's V12 – simply to match the combined firepower of the Testarossa's aerodynamics and its flat-12 engine's power and torque outputs. And there was every reason to re-evaluate this option – contemporary Formula 1 races, and World Rally championships, were being dominated by turbocharged cars. But Alfieri's objective re-examination came to the same conclusion as previously: turbocharging the Countach was a poor engineering solution. The Countach had too little engine bay space for an intercooler and its ancillaries, engine cooling would be an almost insurmountable problem, and the additional loads imposed by high boost pressure would adversely affect the V12 engine's reliability. Turbocharging the Countach also carried other disadvantages: turbo lag, a significant weight penalty due to the mass of the turbo, intercooler and ancillaries, and a muting of the Countach's distinctive and glorious V12's induction and exhaust noises. Alfieri yet again eschewed turbocharging, convinced that he could come up with a better solution to Lamborghini's Testarossa-induced headache.

This time, Alfieri's redesign of Lamborghini's V12 was even more comprehensive. He introduced four valves per cylinder technology to Sant'Agata's production cars, employed down-draught carburettors, redesigned the internal architecture of the V12 with pent roof combustion chambers, and increased the engine's compression ratio as well as its displacement.

The most straightforward of these modifications was increasing the V12's displacement to 5,167cc – by retaining the cylinder bore at 85.5mm but increasing the stroke from 69mm to 75mm. Alfieri already had considerable experience increasing the displacement of this V12 engine having designed a 7.2-litre version for the Class 1 World Offshore Powerboat Racing Championship, and a 7.0-litre version for a modified LM002 off-roader called the LM004 700. Alfieri felt that the engine bore could not be increased safely beyond 75mm, and so decided to pursue a path that he had previously, in 1982, dismissed as 'a useless modification for the Countach': four valves per cylinder technology.

Four valves per cylinder technology would be the cornerstone and the defining feature of this latest Countach variant – so much so that Sant'Agata named this version the Quattrovalvole. Like for like, a 48-valve V12 engine offers better breathing, a higher rev limit, and more power, than a 24-valve V12 engine. This tied in neatly with Alfieri's decision to concurrently increase this new engine's cubic capacity to 5,167cc. In general an engine's torque output increases in an almost linear relationship with its displacement, so by designing a 48-valve, larger-capacity engine, Alfieri easily met both his power and torque objectives.

But there were also other pressing issues that this latest Countach engine would have to satisfy if it was to be homologated for sale worldwide. Air pollution was becoming an ever more urgent and clamorous problem, with environmental activists becoming ever more

Left: A leather-fetishist's delight – an 88½'s sumptuous cockpit.

Opposite: The 88½'s exterior aesthetic was defined by its extended, straked and perforated side-skirts.

lamborghini
countach
F920 OYR

vocal and strident, and governments in some of Lamborghini's most lucrative markets passing ever more stringent legislation limiting noxious gas emissions. Happily, quad-valve-per-cylinder technology increases both an engine's volumetric efficiency as well as its thermodynamic efficiency, thus making the engine more fuel efficient and less polluting. It achieves this by increasing the total inlet and outlet valve areas – so enhancing the inflow of ambient air into the engine's combustion chambers, while simultaneously allowing the exhaust gases to more easily leave the combustion chambers. Alfieri was also acutely aware that quad-valve-per-cylinder technology offers other dividends. Each inlet and exhaust valve could now be smaller and lighter, thereby reducing reciprocating mass. Lower reciprocating mass makes it easier to balance an engine's primary and secondary forces, so making for a smoother engine that is free of damaging vibrations. Lower reciprocating mass also allows for better cylinder head cooling, and for a higher rev limit, both of which can potentially be exploited to generate more power.

Alfieri capitalised on every one of these advantages, but he was still not satisfied. He then realised that he could further improve the Quattrovalvole engine's power and torque outputs, its volumetric efficiency, its thermodynamic efficiency, its fuel consumption, and its noxious gas emissions, by further modifying the inflow of air and fuel into the combustion chambers, and by altering the way this explosive mixture combusted once within the chamber. The twin-choke, centrally mounted, down-draught Weber 44 DCNF carburettors that Alfieri now adopted for the Quattrovalvole engine in place of the previous side-draught carburettors allowed fuel to flow in more freely and in larger volumes. Furthermore, Alfieri redesigned this engine's cylinder head for cross-flow breathing by placing its inlet and exhaust valves on opposite sides. Meanwhile, the new pent roof combustion chambers made space for a centrally placed spark-plug, which allowed for a faster burn time and a more complete burn of the air-fuel mixture. And by choosing to increase the Quattrovalvole engine's compression ratio to 9.5:1, Alfieri ensured that his new power plant produced even more power and torque twinned with better fuel consumption and less noxious emissions.

The Quattrovalvole's new sextet of down-draught carburettors were unusually tall and demanded a central hump in the engine compartment bonnet to accommodate them. Alfieri fashioned this hump in the style of a Mille Feuille puff pastry, which incidentally, like the Countach, is actually of Italian and not French origin. This central hump immediately became a signature feature of the Quattrovalvole. It also resembled a Second World War pillbox, and was therefore alternately referred to as either the Mille-Feuille or the Pillbox. For the first time in a Countach, the engine compartment lid and the front compartment lid were now partly made of Kevlar. Other aesthetic changes however were very minor – slight alterations to the glovebox, central console, facia, and rear panel logo.

To better deploy the Quattrovalvole's new-found power and torque, Sant'Agata upgraded the Countach's tyres to the Pirelli P7R version, and increased the front tyre size to 225/50 VR15 while keeping the rear tyre size unchanged at 345/35 VR15. The only other notable change to the Quattrovalvole's drivetrain over its predecessor's was a new ZF synchromesh system in place of the previous Porsche-designed system.

The Countach 5000 Quattrovalvole comprehensively eclipsed the Testarossa on every performance metric: 455bhp at 7,000rpm, 369lb-ft at 5,200rpm, 183mph, and 4.9 seconds for the 0–60mph dash, versus the Ferrari's 390bhp, 354lb-ft, 180mph, and 5.8 seconds. The figures we quote here for the acceleration times are from independent tests conducted by the highly respected *Autocar* magazine, and are in keeping with the Countach's lower kerb weight at 1,446kg versus the Testarossa's relatively lardy 1,510kg. Lamborghini hailed the new Quattrovalvole as 'the world's fastest production car' at its launch in March 1985. The crown was still in Sant' Agata's hands – for now.

THE COUNTACH EVOLUZIONE

Giulio Alfieri recognised that simply upping the Countach's power and torque outputs would only keep it at the top of the supercar tree for so long. With rival manufacturers developing brand new supercars, which in the early 1980s Sant'Agata simply couldn't afford to do, Lamborghini had to look beyond the Countach's engine to keep its flagship in contention. So Alfieri started to look at the opposite side of the performance-enhancing coin – reducing weight.

Sir Isaac Newton's *Philosophiae Naturalis Principia Mathematica* was published on 5 July 1687, and would in its Second Law of Motion explain the cardinal importance of mass in determining an object's potential acceleration. The equation F=ma, which can otherwise be expressed as a=F/m (where a is the acceleration, F is the vector force applied, and m is the weight) shows that any object can be accelerated faster either by applying a greater vector force to it (in this case by increasing the Countach's torque output), or by reducing mass (in this case by reducing the Countach's weight).

In September 1982 Alfieri started on his twin quests of improving the Countach's acceleration by reducing the car's weight, and of increasing the Countach's top speed by reducing its CdA (aerodynamic drag coefficient multiplied by frontal area). To investigate how these improvements could be wrought he initiated the Countach Evoluzione project. In anticipation of this, he had already sought and obtained funding from the European Economic Community in early 1982 to investigate the use of very lightweight but very strong composite materials in the automotive industry. And in the late autumn of 1982 he employed an aeronautical engineer by the name of Rosario Vizzini to design the Countach Evoluzione.

The Evoluzione was a one-off mobile test laboratory that would allow Alfieri and Vizzini to research how novel materials like carbon fibre, Kevlar, and Nomex, and new aerodynamic appendages like extended side sills and wheel covers, could optimise the performance of forthcoming Lamborghini cars for any given power or torque output. The Evoluzione was never intended to become a concept car, or a prototype car, let alone a production car – it was purely to be a test platform.

At the heart of the Evoluzione lay its carbon fibre monocoque – gone was the Marchesi hand-welded tubular spaceframe chassis. Composite materials were used for all of the Evoluzione's body panels except for its guillotine doors and front and rear wings, which continued to be fashioned from aluminium. Lamborghini claimed that the Evoluzione weighed 1,189kg, but Hubert Hahne, Lamborghini's long-time German importer, independently weighed it and found that it was actually 10kg lighter than claimed. For comparison, *Autocar* found that a regular production 5000 QV weighed 1,446kg.

To improve on the production Countach's dire aerodynamics the Evoluzione featured a new, deeper, lower, and more aggressive front bumper with an inbuilt air dam bearing four Carello spotlights, new composite wheel arches, new extended and flat-sided carbon fibre side sills with perforated strakes at the rear, and new flat, disc-like, carbon fibre wheel trims. The Evoluzione still looked like a Countach, but an unusual and rough-hewn one.

Horacio Pagani, as Head of Lamborghini's Composites Department, was intimately involved in the design, testing, and development of the Evoluzione. In particular Pagani is credited with the Evoluzione's distinctive flat, extended, straked, side sills – a feature that would, in an altered form, find its way on to the 88 ½ and 25th Anniversary Countachs.

This test hack was never painted, giving it its distinctive two-tone, black and silver look. Its body panels were unfinished with ripples, dents, exposed rivets, and tacked-on appendages. Its interior was stark, with only the most necessary sensors and test instruments, and spartan, with absolutely no luxuries, lacking even a sliver of sound-deadening material.

The few available accounts of piloting an Evoluzione paint a picture of a very visceral experience – the car's low mass giving it an immediacy of response to the throttle and brake pedals that no production Countach could begin to match, noises and vibrations that no paying Lamborghini customer would ever accept, and performance in a straight line and around corners that other Countachs could only dream of. The Evoluzione was used to test and evaluate ABS brakes, electronically controlled semi-active suspension, electronically controlled ride height, four-wheel drive, and retractable windscreen wipers.

The Evoluzione carried a special blueprinted engine built by Sant'Agata's Experimental Division, and its quoted power output varied between 480bhp and 500bhp depending on which source you choose to believe. Lamborghini quoted a top speed of 198.8mph and a 0 to 100km/h sprint time of 4.2 seconds for the Evoluzione – figures that were never bettered by any production or experimental Countach. But the Evoluzione's most prized asset was not its performance but the insight – specifically the cutting-edge technical, composite material, and aerodynamic knowledge – it gave to Sant'Agata's engineers.

The unique Evoluzione was destroyed in a planned test crash. Lessons learned from the Evoluzione would go on to inform the later Countach variants, the Diablo, and the Murcielago, and its far-reaching and long-lasting ripples can still be seen and felt in today's Lamborghini flagship, the Revuelto.

THE 88½ COUNTACH 5000 QUATTROVALVOLE

The 88½ Countach can reasonably be called the ultimate production Countach as it benefitted from a unique combination of the most advanced and developed Countach drivetrain, and all the knowledge gained from the Countach Evoluzione test platform, while still retaining Marcello Gandini's glorious, unsullied, almost-original styling. The 88½'s predecessors were denied some or all of these first two ingredients, while the 88½'s successor, the 25th Anniversary, wore a new ballgown of debatable merit, which even today remains a topic of controversy.

Furthermore, by the time the 88½ variant was being put into production, Chrysler was already in full control of Lamborghini, and was insisting on a level of quality control that Sant'Agata had never previously been remotely capable of. Chrysler made this possible by opening its coffers and giving Sant'Agata an unprecedented amount of financial support. No longer would the production line workers have to hurriedly alter components, or modify the way they constructed a particular Countach, just because a supplier had not been paid and a component was therefore unavailable. Lots of Chrysler dollars meant that such compromises no longer had to be made when building the world's finest supercar.

The 88½ variant was the shortest-lived production Countach and probably the rarest. Production of the 88½ started in the last months of 1987 and continued until about August 1988. It is almost certainly even rarer than the LP 400 with a production volume of about 145 units versus 150 for its illustrious ancestor. Only fourteen right-hand drive 88½ Countachs were ever made.

A 25th Anniversary Countach – Horacio Pagani's reworking of Gandini's masterpiece received mixed reviews.

The 88 ½'s signature feature was its Horacio Pagani-designed extended, linear, straked and perforated side sills. These were a development of the radical new sills first seen on the Countach Evoluzione test bed car. Some people loathed these new sills and lamented the passing of the Countach's smooth 'dolphin belly' that Gandini had designed way back in 1971. But many more people loved these aggressive new side sills, which were in keeping with the Countach's masculine image. They were also of the period – Koenig's body kit for the Ferrari BB 512 and the Testarossa, and Strosek's aftermarket makeover of the Porsche 928, all boasted linear strakes somewhere in or on the bodywork – and Lamborghini couldn't afford not to have this most necessary of mid-1980s styling excess. But these new sills were more than just form, they had function too. They increased the Countach's total under-surface area, and thereby, as per Bernoulli's Principle, reduced lift at the rear of the car. Furthermore, the sill perforations directed cool ambient temperature air into the engine compartment and over the rear brake discs and pads, thus improving engine component reliability and longevity as well as braking performance.

The 88 ½'s engine was exactly the same as its predecessor's – an all-aluminium alloy, quad-camshaft, four valves per cylinder, 5,167cc, 60-degree V12. Chrysler insisted that every engine had to prove itself on the dynamometer before being installed into an 88 ½ chassis – and nothing less than 455bhp at 7,000rpm, as well as 369lb-ft torque at 5,400rpm, would do.

Over a leisurely lunch at the Ristorante Giardinetto in San Giovanni in Persiceto on 4 July 2013 Valentino Balboni himself confirmed to the authors that the 88 ½'s front and rear suspension geometries were revised to further improve the Countach's dynamics. Valentino explained that 'the Countach's engine is a little too high', which 'makes it want to oversteer'. The Countach is therefore 'always a little bit dancing', which means that 'the driver has to control the car, and drive with a little bit of feeling'. But Valentino elaborated that it is this flaw that gives the Countach character: 'Constant challenge of controlling the Countach was one reason it is so pleasant to drive.' But not many, if any, of Lamborghini's clientele are as talented as Valentino, and Sant'Agata felt duty bound to set the 88 ½'s suspension geometry for an iota of understeer at the outer envelope of the 88 ½'s cornering potential. This was to warn overconfident but undertalented owners that the car was about to start on what could quickly develop into an uncatchable spin. And in doing this Valentino and his colleagues felt that they had succeeded in making the 88 ½ safer and more drivable than any of its predecessors.

Other changes for the 88 ½ variant were relatively minor. The clutch pedal linkage was modified, making for a lighter clutch action; the rear spring rates were changed; the steering rack was moved slightly; the drivetrain cooling system was revised; and the air conditioning system was improved. The 88 ½ variant also introduced central locking of the guillotine doors.

One important myth that needs to be put to rest is that an electronic climate-control panel with a digital read-out is emblematic of an 88 ½ Countach, and that a straked Countach without this digital display is not a genuine 88 ½ Countach. This is categorically untrue. While many 88 ½ Countachs exported to the United States did feature a digital display, almost all (if not all) right-hand drive 88 ½ Countachs featured the traditional air conditioning controls with a lever, two knobs, and no digital display.

Sant'Agata claimed a top speed of 182mph and a 0–100km/h time of 4.9 seconds for the 88 ½ – figures identical to what the factory had claimed for the Quattrovalvole in 1985. No, what was unique and special about the 88 ½ was threefold: firstly, it successfully married the latest engineering and aerodynamic refinements with maestro Gandini's undefiled, masterpiece design; secondly, it was built to quality control standards that Sant'Agata had never previously entertained; and thirdly, it boasted better roadholding and safer handling than any of its predecessors. The 88 ½ was arguably the most rounded of the production Countachs.

THE 25TH ANNIVERSARY COUNTACH

The 25th Anniversary Countach remains the most controversial of its family line. Undoubtedly the most mechanically developed and comfort-focussed of the production Countachs, this variant suffered, or benefitted, according to your individual viewpoint, from Horacio Pagani's extensive reworking of Marcello Gandini's design masterpiece.

The 25th Anniversary was heavily influenced by Lamborghini's paymaster, Lee Iacocca, the American-Italian boss of Chrysler. And by American customers' expectations – the lucrative American market being key to Lamborghini's financial well-being. Chrysler had initially hoped to launch an entirely new model in time for Sant'Agata's 25th birthday in 1988. But it quickly became clear that this timetable – Chrysler had only acquired Lamborghini on 23 April 1987 – was an unfeasibly ambitious timeframe in which to develop and put into production a worthy Countach successor. The ever-pragmatic Iacocca compromised by asking Sant'Agata's engineers and designers to instead produce a redesigned Countach in time for the company's Silver Jubilee.

Iacocca had spent handsomely to acquire Lamborghini, and then had to bankroll a research and development programme for Sant'Agata's next flagship supercar. Iacocca wanted due recompense for all this expense – he was determined that Chrysler should at the very least be allowed to bask in some of Lamborghini's reflected glory. To this end Chrysler heavily advertised and promoted the links and ties between this Detroit-based mass automobile manufacturer and the Sant'Agata-based purveyor of bespoke supercars. Customers had by now grown accustomed to the Countach's other-worldly styling and Iacocca felt that a substantial facelift could spike fresh interest and generate new sales. American customers were also now becoming an ever bigger and more important segment of Lamborghini's clientele. This was because fuel injection now enabled the Countach to meet strict American and Californian emissions and air pollution legislation, and because the factory and grey market importers had developed bodywork addenda that now met United States Department of Transport regulations on driver, passenger, and pedestrian safety. But these bodywork modifications were often ungainly and appeared like tacked-on afterthoughts, which in fact was exactly what they were. Iacocca told Sant'Agata to make its latest, Chrysler-financed, facelifted, Countach variant not only wholly compliant with current and forthcoming United States legislation, but also to make it more sleek, better integrated, and more homogenous, such that it more closely resembled the arrow-like prototype LP 500 from 1971. Furthermore, Iacocca told Sant'Agata that this last production Countach should be more accommodating of American customers, who tended to be larger in size and less forgiving in nature – these people wanted more cockpit space and more in-cabin luxuries. Meanwhile, Lamborghini's engineers felt that the Countach's handling and roadholding could be further improved, so time and money was also spent refining the car's suspension hardware and geometry.

The 25th Anniversary's restyling was led by Pagani, and he designed a new front bumper-cum-air dam that integrated neatly with the Countach's bonnet. This latest variant retained the 88 ½ Countach's extended, linear, straked, and perforated side sills, but these now came as a one-piece unit together with the rear wheel arch and were of a slightly different pattern. The elephant ear RAM air intakes overlying the rear flanks were replaced by more aerodynamic and less protuberant ducts, but many enthusiasts felt that these were even less elegant than the elephant ear ones. The hot air exit vents behind these intake ducts were also redesigned, but their aesthetic value was questionable. The 25th Anniversary featured a huge, one-piece, rear bumper, completely reworked tail lights, and a new engine compartment lid. And finally and sadly, one of the Countach's signature design features – its NACA ducts – were now made much less conspicuous by having its inner surfaces painted in body colour rather than in trademark black. Such were the dubious improvements that Pagani wrought over the Maestro's fantastical original.

But as per Iacocca's instructions, Americans were better catered for in the 25th Anniversary's cockpit. The glorious-looking but fixed banana-shaped hammock seats were replaced by two-piece electric chairs adjustable for height, reach and squab angle. To meet customer expectations the caricature hand-wound side windows were jettisoned in favour of motorised windows that demanded for nothing more strenuous than pressing down on an electric switch. In the same vein, the cabin temperature could now be controlled effortlessly by push button controls rather than the labour-intensive action of twisting a knob or pushing a tiny lever up or down. The 25th Anniversary's cabin boasted a golf club-shaped gear knob, a leather shield covering the steering wheel boss, redesigned door cards, and right-angled sill covers – every one of which was, and remains, an acquired taste.

Sant'Agata's development of the Countach's dynamics was altogether more successful and welcome. The 25th Anniversary used the stalwart Quattrovalvole engine, and Lamborghini officially stuck to the same power and torque figures that it had first quoted for this engine way back at its launch in 1985 – 455bhp and 369lb-ft. But the unofficial rumour was that the factory had liberated a further 10bhp from this powertrain in the intervening three years.

The 25th Anniversary's trump card was its handling and roadholding, and it had Sandro Munari, one-time Alfa Romeo Grand Prix Team Manager and multiple-time Rally Championship winner, to thank for this. And the ground-breaking Pirelli P Zero tyre.

Lamborghini knew that even its 88 ½ Countach variant suffered from excessive oversteer at high cornering speeds, and could spin suddenly, violently, and without warning, once its limits of adhesion had been exceeded. Munari was employed by Lamborghini in 1987 with the explicit brief to tame the Countach's wayward handling at the extreme margins of its performance envelope.

It was fortuitous that the P Zero tyre, which was launched in 1986, was available to Munari, and he built his revised suspension geometry recommendations based on this new, lightweight, Kevlar-reinforced tyre. Despite the tyre sizes remaining unchanged at 225/50 ZR 15 for the fronts and 345/35 ZR 15 for the rears, the suspension geometry changes that Munari expertly introduced were able to subdue the above-mentioned terminal oversteer and also engineer in some additional steering feel to warn the driver of an impending calamity. Munari was rightfully proud of his achievements and declared the 25th Anniversary 'Sincera' – which translates to, genuine, truthful, and honest to its driver – in a way that none of its predecessors had been.

With its purported 10 extra bhp and its better aerodynamics, Sant'Agata was probably correct in proclaiming the 25th Anniversary variant as the fastest production Countach ever. It was certainly the Countach least likely to bite the hand that guided its steering wheel, and it was also the most cosseting of Lamborghini's flagship cars to date. Many customers loved the Anniversary's new interior and exterior aesthetics, while others were prepared to overlook any desecration of Marcello Gandini's masterful styling in the face of the 25th Anniversary's improved and more trustworthy dynamics. As a consequence, the 25th Anniversary was produced in greater numbers than any other Countach variant – 657 cars had rolled off Lamborghini's Linea 1 production line by the time the Countach's long reign came to an end on 4 July 1990.

CHAPTER 4

THE DEVELOPMENT OF THE DIABLO AND THE EVOLUTION OF ITS MODEL VARIANTS

The Lamborghini Diablo had a difficult conception and gestation. In the earliest stages of its conception there was a small group of Sant'Agata engineers who believed that the Countach could be re-engineered further to keep it relevant into the last decade of the twentieth century and beyond, and they resisted the idea of a new model. When it became amply clear that only a completely new model could meet forthcoming government legislation, customer expectations, societal pressures, and far and fast advancing technological developments that rival manufacturers were already adopting, the fledgling Diablo concept car and then the Diablo pre-production cars became the victims of a tussle between the esteemed Marcello Gandini – arguably the most imaginative, influential, and successful supercar designer of his era – and some relatively unknown Chrysler designers of questionable pedigree who were eager to bask in some of Gandini's reflected glory and Lamborghini's limelight.

PROGETTO 132

The brief: design the world's fastest series production car, with a minimum top speed of 315km/h.

The answer: Progetto 132 – which would go on to become the Lamborghini Diablo.

A Countach successor was mooted in early 1985 by Lamborghini's senior management. The Countach Evoluzione mobile test bed (1983–1987), the Countach 7000 concept car (1986), and the Countach Project 150 restyling prototype (1984–1987) would in due course inform the Diablo. But in the first instance these three experimental vehicles would briefly lend weight to the notion that the Countach could be updated, before the realisation dawned on even the most ardent of Countach adherents that this seminal supercar was now so seriously outdated that it could not be evolved yet again into becoming the fittest, and so

Another treasured model – this time of Lamborghini's Progetto 132 prototype.

This sleek, original Diablo shows off Marcello Gandini's trademark slashed and uprising rear wheelarch.

did not deserve to survive. But well before these three prototypes were built and tested, Patrick Mimran already knew that a new Lamborghini flagship was needed – but he did not have the money to finance such a major project. Mimran understood that only a major automotive manufacturer would have the necessary wherewithal, and in any case he was already thinking about selling his Italian supercar company. He also knew that if Lamborghini already had a model designed and ready to be put into series production then the company would be hugely more attractive to a potential buyer.

So in June 1985 Mimran and Emile Novaro, Lamborghini's President, formally initiated Progetto 132 – a study to design and develop Lamborghini's next flagship car. The rear mid-engine format with a longitudinally placed V12 engine, and Paolo Stanzani's brilliant South-North engine-gearbox placement, had proven themselves over the last decade and a half and Lamborghini's engineering team quickly decided that all these features had to be retained for the new car. Four-wheel drive was also imperative. The Porsche 959 prototype had made its debut at the 1983 Frankfurt Motor Show, and by the time customer cars were ready to be delivered in 1985 the 959 had already proven the importance and value of four-wheel drive technology in a supercar by winning the gruelling 1984 Paris–Dakar Rally. Lamborghini could not afford to be left behind, and as luck would have it retaining the principal drivetrain features of the Countach as already outlined above allowed for a relatively cheap and easy conversion from rear-wheel drive to all-wheel drive – but only once Lamborghini had worked through the finer details of making its novel Visco-Traction four-wheel drive system reliable and user-friendly.

While Stuttgart had been working on the Porsche 959, Maranello had been dreaming up the iconic Ferrari F40. The Nicola Materazzi-designed F40 entered production in 1987 and employed state-of-the-art intercooler-enhanced turbo technology as well cutting-edge Kevlar, carbon fibre, and other composite materials. Progetto 132 had a battle on its hands in the form these two formidable adversaries and the advanced technologies that they deployed.

The Countach's guillotine doors had been an instant and perpetual hit with its extrovert owners and bedazzled onlookers alike, and this feature too had to be retained. And who better to design Progetto 132's body than the man who had not only dreamt up these astounding doors, but who had also previously proven himself to Sant'Agata's management by designing the Miura and the Countach. Marcello Gandini, the foremost supercar stylist of his era, knew that Progetto 132 had to be much more aerodynamic than the Countach to meet its most important design briefs – higher performance, better fuel consumption, and lower noxious emissions. Key markets – the United States, Northern European countries, and Japan had already passed ever-more stringent legislation on petrol consumption and tail-pipe pollutants and many other countries were threatening to follow suit. Progetto 132's body had to be dramatic yet beautiful, while Lamborghini's ever more demanding and ever wider-girthed customers wanted better air conditioning, better passive safety features, better forward, side, and rear visibility, easier ingress and egress over the wide sills and through the narrow guillotine door apertures, more headroom, more legroom, and larger and more comfortable seats.

Luigi Marmiroli was appointed Lamborghini's Technical Director in 1985, and he was tasked to work with Gandini to secure the aerodynamic, top speed, petrol consumption, and emissions targets detailed above. He was also tasked with making Progetto 132 more user friendly than its predecessor with better handling and roadholding, lower noise, vibration and harshness levels, a softer clutch, an easier gear change, and less violent and more forgiving responses once the admittedly very high limits of its performance envelope had been breached.

An octet of engine compartment bonnet vents and a quartet of exhaust pipes give the rear of this original Diablo real presence. Note the uninterrupted sweeping curve from tail to nose.

The cockpit of an original Diablo.

An American-spec original Diablo. Note the prominent front spoiler, the quadrilateral sill air-intake, and the protruding rear bumper.

While Gandini honed Progetto 132's wedge-shaped body to minimise aerodynamic drag, Marmiroli reworked Lamborghini's Bizzarrini V12. In early March 1987 the pair presented their first prototype – Chassis 12000 – to Lamborghini's Board members. This car was fitted with the 5.2-litre Type 521 engine and in late March it began extensive wind tunnel testing. The resulting data was hugely heartening – Chassis 12000 cleaved through the air with great aerodynamic efficiency. The Diablo was on its way to production.

Or maybe not. In April 1987, Lee Iacocca, the President of Chrysler bought Lamborghini from the Mimran brothers for a rumoured US $25 million. He had promised Lamborghini its independence while guaranteeing it the necessary financial, technological, and logistical support to make it the pre-eminent supercar manufacturer of the day. Chassis 12000 was meanwhile being put through its paces on real-life Italian roads by Valentino Balboni, and Valentino's feedback mirrored and supported Progetto 132's excellent wind tunnel data. Marmiroli, Gandini and Novaro were delighted.

On 1 June 1987 Iacocca visited Sant'Agata for the first time. He inspected Chassis 12000 carefully, and pronounced it 'Terrific'. But shortly thereafter, Chrysler's middle management started calling for 'minor modifications' to Chassis 12000 that could only have been authorised by Chrysler's top executives. Minor modifications soon became a Chrysler-led call for a substantial reworking of Progetto 132, leaving Marmiroli and Gandini in despair as they felt that their prototype was almost ready to go into series production. Gandini was particularly offended by Tom Gale and his fellow Chrysler designers, who appeared to regard Progetto 132 as a way to make their names known in the car design world. These almost unknown Chrysler stylists submitted their own proposals for a radically different Progetto car, but their designs were of questionable merit. Gandini rejected their proposals but agreed to rework his own design to make it more to Iacocca's liking. Gandini's second offering was smoother and without the sharp edges that characterised his first proposal. Crucially this reworking remained just as aerodynamic as Chassis 12000. But Iacocca was still not happy, particularly with this second car's rear end, which he condemned as being 'too busy'. In June 1988 Gandini put forward his third Diablo styling proposal, and yet again Tom Gale and his Chrysler colleagues mistakenly thought that they could better the Maestro's best efforts. Novaro now confronted Iacocca and demanded that Lamborghini be allowed the independence that it had been promised. Novaro pointed out that the Countach would imminently lose its homologation rights in several key markets, leaving Lamborghini with no cars to sell. Financial catastrophe beckoned unless Chrysler allowed Gandini's third proposal to go into series production. Faced with Chrysler's profit margins being severely squeezed, Iacocca relented, while Novaro and Gandini also compromised by giving Chrysler's William A. Dayton almost free licence to redesign the new car's cockpit and interior.

Antonella Bertuzzi developed Progetto 132's tubular spaceframe chassis using computer-assisted analysis and the resulting structure used far fewer box sections. Notably this spaceframe was constructed from square tubes instead of the round tubes used in the Countach. Round tube spaceframes are the ultimate – but are more difficult and more time-consuming to weld. Lamborghini wanted to sell the Diablo in greater numbers than the Countach, and spaceframe chassis production had been a key limiting factor in Countach output, so it made sense for the Diablo to use a spaceframe that required less expert welders and which could be welded together more quickly.

Meanwhile, Marmiroli rejected the 5.2-litre Type 521 engine even though it had proven itself reliable, flexible, and capable of producing 470bhp. He did this for two important reasons. Firstly, Marmiroli was

This original Diablo boasts spartan-looking but surprisingly comfortable hammock-style seats.

Aesthetics-wise, the four-wheel drive Diablo VT is pretty much indistinguishable from its predecessor, the two-wheel drive original Diablo.

uncertain that the Type 521 engine could provide the power and torque needed to breach the 315km/h top speed design brief, or to post world-beating acceleration times. Secondly, Sant'Agata's rivals were now producing far more powerful engines and Lamborghini had to keep up with its challengers.

Luigi Marmiroli's completely new Type 522 engine (detailed below) was installed into a brand-new bodyshell based on Marcello Gandini's third design proposal. This car – Chassis 12001 – can be considered the first Lamborghini Diablo.

THE ORIGINAL DIABLO

Sant'Agata's successor to the Countach was unveiled to the public at the Sporting Club de Monaco on 21 January 1990. The Diablo was the last Lamborghini supercar conceived and launched while the company's founder Ferruccio Lamborghini was still alive. At its debut, Sant'Agata proudly claimed that the Diablo, just like its illustrious predecessors the Miura and the Countach, was the fastest series production car of its time. Beautiful in appearance, with a mechanical specification to match, and an unrivalled pedigree, Novaro and Iacocca were delighted, but not completely surprised, that enough new orders flowed in over the course of the Diablo's two-day Monaco launch to fill the order book for the next year.

The Diablo's styling was an evolutionary refinement of the revolutionary Countach design that had stunned the supercar world in 1971. The Diablo was wedge-shaped, wide, low, long, featured Gandini's calling-card slashed and uprising rear wheel arches, had vents galore, and of course boasted guillotine doors and retractable headlights. But in the name of aerodynamic efficiency the Countach's origami-straight lines and acute bends were jettisoned in favour of smooth surfaces and drag-eschewing curves. These changes alone lent the Diablo a more modern look. One of the Diablo's signature features was its drooping side windows that appeared to start in the middle of the car before falling dramatically forwards and downwards almost to the nose.

Unlike the Countach, whose body was made almost exclusively of expensive and time-consuming to fashion, hand-beaten aluminium panels, the Diablo's body was made of a variety of materials. Importantly, its alloy panels were machine stamped – which was much cheaper and quicker than having expert artisans work individual metal sheets with a hammer and a buck. Autoclave, a composite material containing Kevlar and fibreglass, was used for the front and rear bumpers, side panels, entire lower body, and front and rear bonnets. The guillotine doors and front and rear wings were made of aluminium alloy, while the all-important passenger safety cell was constructed from high and medium-strength steel and a variety of high-strength alloys. Sant'Agata was happy to pay for extra-strong but lightweight materials wherever necessary because a low kerb weight was crucial for meeting the Diablo's fuel efficiency, noxious emissions, and acceleration targets. The Diablo's Vaccari & Bose-fabricated square-section tubular spaceframe chassis weighed only 80kg – a reduction of 30kg over the round-section Countach's Marchesi-fabricated spaceframe. Gandini buried a myriad of ducts and vents deep into the Diablo's smooth bodywork so as to present a clean and aerodynamic frontal aspect to the oncoming airstream while still allowing in lots of ambient temperature air to cool its massive Bizzarrini V12 engine. Six engine bonnet vents enabled hot air to readily escape the engine compartment. Silver-coloured 17in, five-holed, light alloy wheels dominated the Diablo's arrow-like profile, while two massive 335 profile rear tyres, four tail lights, a quartet of bazooka-like tailpipes, and an octet of rear vents dominated the car's rear.

The Diablo SE30 boasted an even more arrow-like profile with a modified front bumper-cum-air dam. Note the altered sill air-intake, now with two almost vertical blades.

The limited edition, ultra-exclusive, Diablo SE30 is seen here in its most iconic colour. Note the wreathed SE30 anniversary badge to the left of the silver Raging Bull emblem.

The Diablo SV's moniker stands for Super Veloce, and Sant' Agata had only ever used it previously on the hallowed Miura SV.

The Diablo SV was targeted at the driving enthusiast and sacrificed top speed for the ultimate in acceleration, braking and cornering potential.

The Diablo's cabin was styled by William Dayton and majored on comfort and safety in a way that Countach owners could only dream of. Fine, hand-stitched leather covered just about every visible surface. The Diablo's door apertures were more generous than the Countach's, allowing for easier entry and egress, and the foot pedals were now spaced slightly further apart. The greater glass area gave the Diablo a lighter and more airy feel, and the thinly padded one-piece hammock seats were more cosseting. A steeply ascending central console column with a golf club-shaped gear lever at its base housed the side window switches, an Alpine hi-fi system's controls, the buttons operating the parking, side, interior, and fog lights, the fuel filler cap release, and the heating and ventilation buttons for the car's climate control system. Ergonomics had moved apace since the Countach LP 500 prototype's debut and Dayton designed the Diablo's primary and secondary controls to be much more accessible and user-friendly. The only exception to this rule was the Professor Renato Meduri-designed multi-tier, angulation-adjustable, dashboard with the speedometer and the rev counter at the extreme left and right corners of the dashboard's top storey. Meduri's rationale was that this arrangement would allow the driver to view both dials through his peripheral vision while keeping his or her central vision focussed on the road ahead. However, any half-baked eye surgeon could have told Meduri that this would not work in practice, as seeing the dial needle tips requires central vision resolution rather than peripheral vision resolution. Additionally, locating the dials where Meduri chose to place them required two discrete and time-consuming eye movements, whereas a conventional placement would only have required one eye movement. Needless to say, Lamborghini abandoned this bizarre arrangement at the first opportunity and have never returned to it since.

The Type 522 engine that Marmiroli developed for the Diablo was a variant of the classic 60-degree Bizzarrini V12. This 5,707cc all-alloy, quad camshaft, naturally aspirated, four valves per cylinder, wet sump powerplant had a bore of 87mm and a stroke of 80mm. Operating at a compression ratio of 10:1, it produced 492bhp at 6,800rpm and 580Nm torque at 5,200rpm. The 522 engine used timing chains with dampers, an automatic timing tensioner system, and a bespoke Marelli-Weber integrated injection-ignition arrangement that Lamborghini christened LIE (Lamborghini Iniezione Elettronica). Marmiroli felt that the huge amount of torque that the 522 engine produced made a six-speed gearbox entirely surplus to requirements, but a brand-new five-speed gearbox working with a dry single-plate clutch and a ZF limited slip differential was developed for the Diablo.

The Diablo had non-assisted manual steering with three and a half turns from lock to lock, giving a minimum turning circle of 13.2 metres. A twin hydraulic braking system with brake booster assistance and 330mm vented front discs and 284mm vented rear discs provided the necessary stopping power. The suspension incorporated anti-dive and anti-squat technology together with anti-roll bars and single co-axial spring-shock absorber units on the front axle, and twin co-axial spring-shock absorber units on the rear axle. Asymmetrical, tubeless, 245/40 ZR 17 and 335/35 ZR 17 Pirelli P Zero tyres were wrapped around lightweight aluminium alloy OZ wheels measuring 8.5 × 17in and 13 × 17in on the fronts and rears respectively.

With a kerb weight of 1,576kg and a weight distribution of 60% rear and 40% front, the Diablo's fuel consumption figures were eye-watering: 14mpg for the EPA highway cycle dropping down to 9mpg for the EPA City cycle. But all this was forgotten and forgiven when the Diablo was officially clocked at the Nardo test track decimating the 0–100km/h dash in just 4.09 seconds on the way to reaching a top-speed of 325km/h – thereby eclipsing the Ferrari F40, the previous production car top-speed recordholder. Mission accomplished.

The Diablo SV-R was a lightweight derivative of the Diablo SV road car, and was designed for the track.

THE DIABLO VT

The Diablo VT was an evolution of the Diablo with DNA that could easily be traced back to the Countach save for one truly revolutionary innovation that would go on to define the majority of Lamborghini supercars for decades to come. The VT made its debut at the Geneva Motor Show in March 1993 and the radical deviation that it introduced was the unique and state-of-the-art Lamborghini Viscous-Traction four-wheel drive system. Every Countach variant and the original Diablo suffered from a tendency to oversteer, and Giulio Alfieri, Sandro Munari, and Valentino Balboni had tried, with varying degrees of success, to tame this unwanted trait. The increased power and torque of the Diablo would only make oversteer more likely and more pronounced, so the original Diablo was meant to have been launched as an all-wheel drive car. But the novel Viscous Traction system was still being worked on in 1990 and was far from production ready.

Stanzani's revolutionary South-North engine-gearbox system that made its first appearance in the 1971 Countach LP 500 prototype lent itself perfectly to be altered and integrated into a four-wheel drive system. Marmiroli investigated a fully mechanical all-wheel drive system as well as a hydraulic all-wheel drive system but in the end concluded that a part-time system offered the best solution for a Lamborghini flagship supercar. The Viscous Traction system directed the vast majority of the Type 522 engine's torque to the gargantuan 335 section tyres for the vast majority of the time, only directing torque to the front-axle tyres when the rear tyres were about to break traction and spin. When the rear tyres did lose grip the Viscous Traction system would almost instantaneously direct up to 25% of the available torque to the front tyres, thereby allowing these tyres to play their part in averting an accident.

The Diablo's Viscous Traction system consisted of a viscous coupling integrated into the forward-placed gearbox. An additional casing positioned in front of the gearbox housed a carbon fibre driveshaft taking torque to the front limited slip differential and the front driveshafts. The rear differential and rear driveshafts were driven in conventional Countach fashion with a drop-down gear set and an in-sump driveshaft directing torque to the rear of the car.

Two lesser innovations also made their debut in the Diablo VT. Both would go on to feature in many later Lamborghini supercars – power-assisted steering and electronic damping. Early adopters of the Diablo had voiced the same misgivings as many Countach owners about their cars' excessively heavy steering. This problem, and the consequent complaints it generated, only got more pronounced as these cars' front tyres grew in width and the cars grew in weight. The type of customer Chrysler was targeting for the Diablo was very different to the hardy, devil may care, Countach customer, and in-cockpit comfort and greater ease of driving were now key requirements. TRW designed the Diablo VT's power-assisted rack-and-pinion steering system, which boasted 3.2 turns lock to lock and a turning circle of 13 metres. The challenge here was to ensure that this new power-steering system did not excessively filter out the feedback messages coming from the road surface to the driver via the steering column and steering wheel. Nor did Lamborghini want to lose the steering wheel's all-important self-centring as this informs the driver about just how hard a car is cornering. This is particularly important at high speeds, so Lamborghini and TRW settled upon an arrangement where the degree of power assistance decreased with increasing road speed and with the degree of steering wheel angulation away from the straight ahead position. The VT's power-assisted steering system received much-deserved praise from its grateful and relieved owners.

Undecapitated, the Diablo Roadster was hardly a looker.

Losing the roof transformed the ugly duckling undecapitated Diablo Roadster into a stunning swan.

This pre-facelift Diablo Roadster's detachable top sits beautifully flush over the car's engine compartment bonnet.

The third innovation first seen on the Diablo VT was its Koni developed and built suspension system featuring electronically controlled shock absorbers whose damping parameters could be altered remotely. As well as an automatic setting, the driver could manually select between four different levels of shock absorber damping. The rest of the VT's suspension system remained unchanged from the original Diablo's set-up.

The VT gained new four-piston alloy brake callipers made by Brembo, enlarged air dam inlets, a strengthened gear lever, improved cabin ventilation, an upgraded electrical system, and a conventional instrument panel in place of the Meduri-designed, multi-tiered gauge display.

The VT's drivetrain was largely the same as its predecessor's except for a strengthened bottom end, a beefed up valvetrain, and slightly modified ignition timing. The factory claimed that the VT put out exactly the same power and torque outputs as the original Diablo. And despite the VT weighing 47kg more than the original Diablo courtesy of its new and heavy Viscous Traction system, Sant'Agata claimed the same fuel economy and performance figures for the VT as for the original Diablo. The VT represented a major step forward along the Diablo's evolutionary pathway, with one radical innovation, two cutting-edge developments, and a host of smaller but still important refinements.

THE DIABLO SE30

The Diablo SE30 was a limited edition road racer and a 30th birthday present from the Lamborghini factory to itself. The SE part of the SE30 name stood for Special Equipment, while the 30 celebrated the company's Pearl Jubilee. The SE30 was designed from the start to be a highly exclusive Diablo variant – only 150 units would be produced, and each would cost £16,000 more than a regular Diablo VT.

This very special Diablo was designed not only for the Italian Supercar GT Championship in which its principal rivals would be the Ferrari F40 and the Porsche RSR, but also for the road, where it would go head-to-head with the likes of the Bugatti EB 110 and the Jaguar XJ220.

The SE30 would attempt to dominate its road and race rivals through a combination of increased power, reduced weight, new suspension technology, and a sensuously beautiful and much more aerodynamic bodyshell than the Diablo variants that had preceded it.

It is ironic that on the thirtieth anniversary of Ferruccio Lamborghini's inauguration of his eponymous supercar company, Sant'Agata chose to go directly against Ferruccio's founding diktat that Lamborghini should always eschew racing. The SE30 was designed primarily for customers who wanted to go racing. To this end the 5,707cc was modified with larger-diameter inlet valves, a 7kg lighter crankshaft, a more free-flowing exhaust, and a reprogrammed LIE electronic fuel injection map. With these modifications the Type 522 engine now produced an additional 28bhp, taking its peak power output to 520bhp at 7,100rpm. Peak torque remained unchanged at 580Nm but it was now produced at 5,900rpm instead of 5,200rpm.

The SE30 weighed 176kg less than the regular Diablo VT. To achieve this massive weight reduction, Sant'Agata reverted back to two-wheel drive and manual steering (although power-assisted steering could be ordered as an optional extra); replaced the glass side windows with lightweight, fixed Plexiglass panels; used carbon fibre wherever suitable in the interior and on the exterior to save every last gramme; swathed the cockpit in Alcantara instead of leather, which is heavier; employed racing buckets instead of sumptuous seats; and jettisoned the high-fidelity sound system and air conditioning (although air conditioning

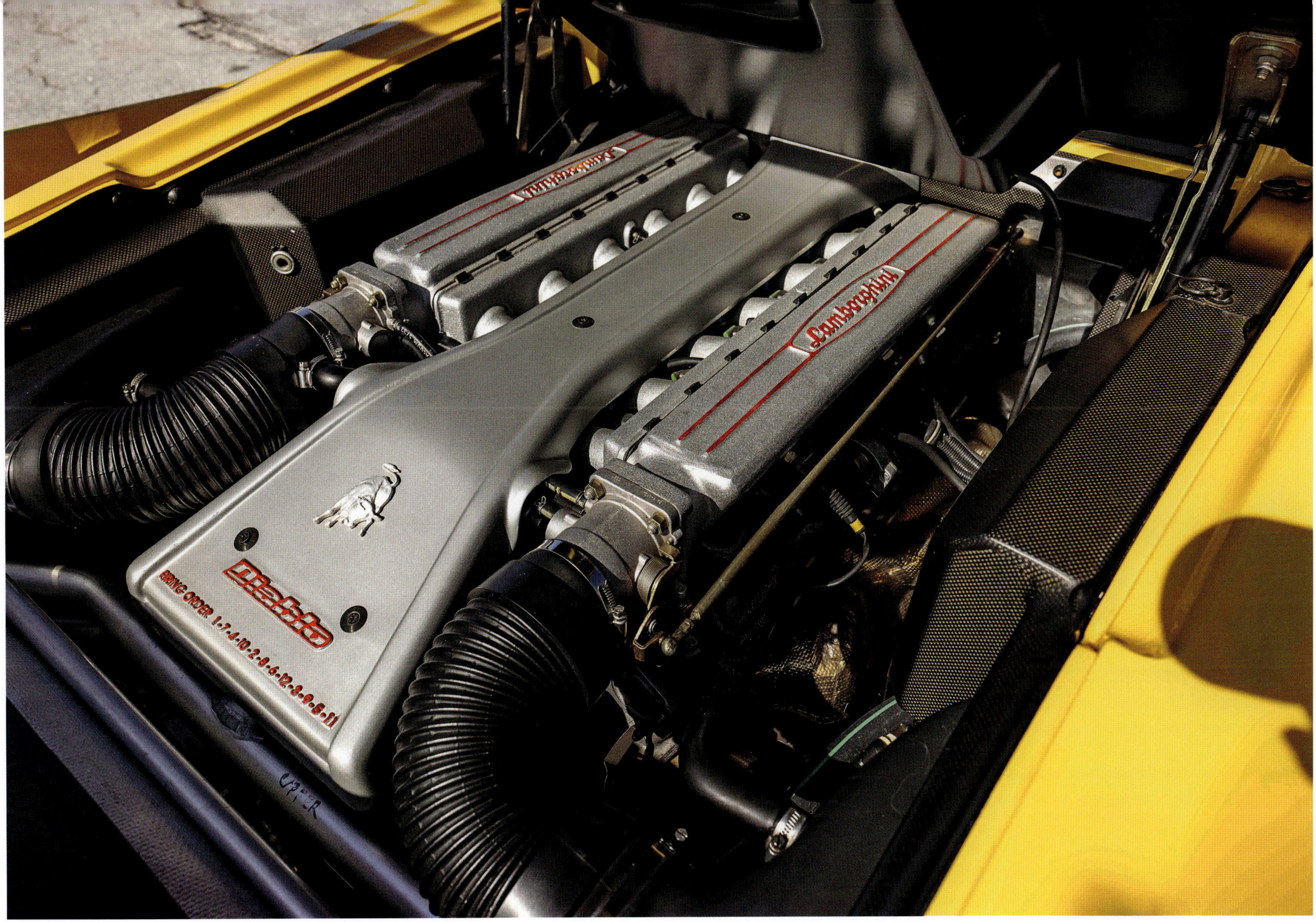

Note the Firing Order of this Diablo's Bizzarrini V12 engine.

A 2nd generation, 1999 Model Year, Diablo SV. Note the flush-fitting, non-retractable headlights and the twin roof-mounted air-scoops.

could be specified as a no-cost option. Special magnesium wheels and titanium wheel nuts gave a 25% weight reduction over their aluminium equivalents – and as this weight-saving was in unsprung mass it directly benefitted the car's handling and roadholding. Although the SE30's front and rear tyre sizes and front wheel size remained unchanged, the rear wheel size was increased to 13 × 18in.

The SE30 introduced two new suspension technologies that had not featured previously in any series production Lamborghini – traction control and in-cockpit adjustable anti-roll bars. The traction control system could be turned off completely, or the driver could select one of three settings that would determine at what degree of wheelspin the traction control system would intervene and cut fuel flow to the injectors, and thereby reduce the engine's torque output until full traction was regained. The adjustable anti-roll bars could be set to one of three positions, thereby allowing the driver to optimise the car's grip and composure on any given surface. To compensate for this road racer's lower mass, its shock absorber and spring rates were subtly altered.

But the most striking thing about the SE30 was its stunning aesthetics. The SE30 was even more sleek and arrow-like than the VT, and every part of its bodyshell appeared to gel more harmoniously with its neighbouring panel. The front air dam and front spoiler were shaper, more protuberant, and altogether more aggressive, with slatted brake cooling ducts and four sunken-in driving lights, and the black and gold Lamborghini emblem now embedded deep into the air dam. The air scoops immediately ahead of the rear wheels were substantially reworked and would become a signature feature of the SE30, and in time would star in other highly exclusive and hyper-desirable Diablo variants. These new scoops had two diagonal and backward-sloping blades forming three almost vertical air inlet vents. To improve engine cooling and for aesthetics, the engine compartment now featured a Venetian blind-style carbon fibre bonnet a la the Miura. A redesigned rear bumper was complemented by a massive rear wing with an adjustable central tab and downward curving endplates.

Four-point racing harnesses; grippy Alcantara; drilled foot pedals; racing bucket seats; a smaller diameter, flat-bottomed steering wheel; and an optional inbuilt fire extinguisher system all spoke of the SE30's quasi-racing ambitions.

An even more desirable and exclusive sub-variant was the highly coveted Diablo SE30 Jota, which boasted a factory-supplied set of bodywork and mechanical parts. Only twenty-eight Jota upgrade kits were ever built, and since these were specially made by Sant'Agata's sideshoot Lamborghini Engineering S.p.A. – its ex-racing and Formula One division – the Diablo SE30 Jota was not only very rare but also carried extra kudos and provenance.

Weighing in at 1,449kg, and propelled by 520bhp at 7,100rpm and 580Nm at 5,900rpm, the Diablo SE30 could accelerate from rest to 100km/h in 3.9 seconds and reach a top speed of 207mph. But even more bewitching than these performance figures were the SE30's rarity, novel technology and aesthetics.

THE DIABLO SV

Sant'Agata uses the SV moniker sparingly. SV stands for Super Veloce and only the hallowed Miura SV had previously been honoured with this epithet. When the Diablo SV was presented at the 1995 Geneva Motor Show Lamborghini proudly introduced its latest car as: 'The quickest and most responsive road car that Lamborghini has ever produced.'

The early 1990s had been a turbulent time for the world economy, and supercar manufacturers were particularly susceptible to these recessionary pressures. Chrysler had been hit hard and as a consequence had put Lamborghini up for sale. Financial and business challenges abounded globally, and the typical Lamborghini customer was particularly vulnerable. Sant'Agata had, however, identified a gap in the market between its safety and comfort-orientated Diablo VT, and its highly exclusive, race-orientated SE30. The limited edition Diablo SE30 had been a great success and Sant'Agata now sought to capitalise on this by taking some of its best bits and making them available in the series production Diablo SV.

The SV was targeted squarely at driving enthusiasts, and for this model variant Lamborghini was happy to sacrifice top speed for the ultimate in acceleration, braking, and cornering. Superlative responses to driver input and excellent driver feedback were prioritised above all else by the factory. Early on in the SV's conception and development phase Chrysler had sold Lamborghini to Megatech, and this Indonesian-led conglomerate had installed Mike Kimberley – a protégé of Lotus's Colin Chapman – as Chairman. Kimberley was totally invested in the principles behind Chapman's famous maxim: 'Adding power makes you faster on the straights. Subtracting weight makes you faster everywhere.'

Bowing low before the prevailing economic environment, Kimberley knew that the next Diablo variant that Lamborghini launched also needed to be cheaper than its previous offerings. So he decided that the SV would be a lightweight, stripped-out car – cheaper to make and therefore more affordable, and more driver-focussed and more responsive courtesy of its reduced mass.

At 1,530kg the SV was 93kg lighter than the VT. This weight reduction was achieved by employing a two-wheel drive system rather than a four-wheel drive system, and by using composite materials wherever suitable. Surprisingly for a purportedly performance-orientated car in which traction control had been eschewed in the name of weight saving, power steering, air conditioning and even a hi-fi system were all retained! The 5,707cc Bizzarrini engine was further developed to produce an additional 18bhp through the use of more aggressive cams and a more free-breathing exhaust system. Peak outputs of 510bhp at 7,100rpm and 580Nm torque at 5,900rpm were the end results.

The most striking thing about the SV's bodyshell was its two roof-mounted air intakes as per the very rare Diablo SE30 Jota. Huge italicised SV logos splashed across the car's flanks came as standard, but more self-conscious customers could ask for these stickers to be left off. A single-plane rear spoiler with an adjustable central trim tab came as standard, while the front spoiler-cum-air dam was modified, with the Lamborghini emblem now returning to its usual position at the base of the front bonnet.

Larger ventilated front discs made by Brembo, rather fussy new alloy wheels with coloured centres and silver rims, a shorter final drive ratio, and almost no sound-deadening material, rounded off the SV's standard specification. As per the SV's stripped-out, driver-orientated, design brief, it traded top speed (still 187mph) for stunning acceleration (0–100km/h in 3.9 seconds), and its launch price of £125,000 was £19,000 cheaper than the VT's and £50,000 less than the SE30's. As well as the cosmetically modified limited edition SV Monterey Edition, the factory also produced the more consequential Diablo SV-R – a lightweight racing adaptation of the SV road car – for competition in the Lamborghini Phillipe Charriol Super Trophy Championship.

THE DIABLO ROADSTER

The Diablo Roadster was a Lamborghini first – never before had Sant'Agata put an open-top version of its flagship car into series production. Although the Roadster was only launched in December

1995 at the Bologna Motor Show, its genesis dated back to the Chrysler era when Lee Iacocca had strongly championed an open-top version of the Diablo. Iacocca and Lamborghini were both keen to have a presence in every niche of the supercar market, and the lack of a convertible flagship was a gaping hole in Lamborghini's offerings. Marcello Gandini was tasked with designing this convertible and he had presented his 'Lamborghini Diablo Roadster Prototype' at the 1992 Geneva Motor Show. This car was highly impractical, even for a concept car, with an almost non-existent windscreen and a very large, removable, one-piece hard top. Many alterations and modifications were necessary to make this prototype ready for series production but all this hard labour came to fruition with the final three-in-one car that was revealed in Bologna. With its hard-top in place the production Diablo Roadster looked more than a trifle ugly, while with the hard top stashed over the engine compartment bonnet it looked handsome. But with the hard top left in the care of one's butler, the unadorned and unencumbered Diablo Roadster looked truly stunning.

This hard top was made of carbon fibre, and could, albeit with some difficulty, be attached and detached by one person. It was secured over either the cockpit or the engine compartment bonnet by four latches. But once mounted over the engine compartment bonnet there was no way to access these latches (not a problem when mounted over the cockpit) so Sant'Agata devised an ingenious solution – a complex vacuum-operated mechanism released the latch studs when a switch mounted high up on the firewall behind the driver was activated.

To accommodate the detachable hard top, the rake of the Roadster's windscreen had to be altered, the small side windows behind the doors had to be replaced with fixed panels, the engine compartment bonnet had to be substantially restyled, and most prominently the air intakes above the rear flanks had to be completely redesigned and enlarged to compensate for the inevitable reduction in airflow (and therefore engine heat dissipation) when the hard top was secured over the engine compartment bonnet. The Roadster boasted specially commissioned 17inch, five-hole, alloy wheels, and redesigned seats. The cabin leather was specially treated to make it more resistant to ultraviolet light and acidic raindrops.

With the limited edition SE30 no longer in production, the Roadster became the most expensive Diablo variant. Even the same optional extras cost more for a Roadster than for its hard top siblings. The additional bracing needed to compensate for the Roadster's reduced torsional rigidity meant that it weighed more than its fixed-head compatriots at 1,625kg. The Roadster had few other modifications to its chassis and suspension, and almost no alterations to its drivetrain. Its 5,707cc engine produced 492bhp at 7,000rpm and 580Nm at 5,200rpm, allowing it to sprint from rest to 100km/h in 4.0 seconds and reach a top speed of 201mph. Sant'Agata had every justification to unequivocally call its decapitated Diablo the world's fastest four-wheel-drive convertible and the world's fastest production convertible.

The Second Generation VT, Roadster, and SV (1999 Model Year)

Challenged on just about every front by the arrival in 1996 of Ferrari's front-engined 550 Maranello flagship, and without the financial firepower to develop a new model, Sant'Agata responded by facelifting its entire 1999 Model Year Diablo range. No revolutionary changes, just simple but comprehensive evolutionary refinements. Ferrari's F355 junior supercar, which was launched in 1994, was another hugely influential factor here – it was classically beautiful yet also very modern looking, it was packed with technological innovations, and it banished poor ergonomics and unfriendly driving characteristics into the dustbin of history books. Lamborghini had no choice but to respond.

Retractable headlights, once synonymous with Lamborghini theatre, had to be ditched in this facelifted 1999 Model Year Diablo SV because of imminent pedestrian safety legislation.

All three facelifted, second-generation, Model Year 1999 Diablos – the VT, the VT Roadster, and the SV – were launched at the Paris Auto Show in October 1998. The single most striking thing about these second-generation Diablos was their fixed, faired-in, glass-covered headlights, which replaced the retractable headlights that had been a signature feature of every Countach and every previous Diablo variant. A fair bit of theatre was lost through this swap, but fixed headlights allowed the Diablo to conform with forthcoming safety legislation. The US National Highway Safety Administration's Standard No. 108, and the European Union's legislation on 'Car's Front End Deformability', would soon make pop-up headlights essentially illegal. Furthermore, there were good engineering and marketing reasons for abandoning retractable headlights. When elevated they acted like horizontal parachutes, substantially increasing the car's aerodynamic drag coefficient (Cd), and then compounding this by also increasing the car's frontal area (A), and so massively increasing the car's all-important CdA. A poor CdA brings with it poor acceleration, a lower top speed, bad fuel consumption, and higher emissions. Additionally, customers had become blasé to the aesthetics of Lamborghini's pop-up headlights over the preceding twenty-eight years, so the new fixed, faired-in headlights were a novel and tasty morsel with which to tempt potential, previously jaded, clients into opening their wallets. Lamborghini neatly sidestepped the huge cost of homologating new headlights by using Nissan 300ZX Z32 headlights under licence. The other notable exterior change was the adoption of the SE30's modified lower air scoops with its two diagonal and backwards-sloping blades for the facelifted VT and VT Roadster variants – either as standard (USA market) or as an optional extra (other markets). The SV retained the old-style quadrilateral lower air scoops. Later in the production run, two cosmetically modified, limited edition subvariants of just twelve units each were built solely for the USA market – the Alpine Edition Coupe and the Momo Edition Roadster.

For the first time in Lamborghini's history, the factory now fitted driver and passenger airbags as standard. To accommodate the passenger airbag the whole of the Diablo's dashboard had to be redesigned and Sant'Agata took this opportunity to refresh the dash display. Another major safety innovation introduced with the second-generation Diablos was the Lucas Varity/Kelsey Hayes anti-brake locking system. These cars also featured Dynamic Rear Brake Proportioning, which optimally distributes the available braking force between the front and rear axles. Larger front and rear brake discs with new brake callipers demanded larger wheels, so these facelifted Diablos received 18in OZ Racing split rim aluminium alloy wheels. A new, state-of-the-art, electronically controlled, variable valve timing system; a new ECU with full diagnostic and data-acquisition capability; revised cam profiles and larger valves; upgraded catalytic converters and lambda sensors; and a new fuel-injection system, together enabled these latest Diablos to easily conform with even California's draconian noxious pollutant rules that were about to come into force in 1999. Again, all this was driven by legislation, customer expectation, and inter-manufacturer supercar rivalry.

All these second-generation Diablos retained the 5,707cc engine, which now produced 530bhp at 7,100rpm and 605Nm at 5,500rpm. The four-wheel drive VT and VT Roadster both weighed in at 1,625kg. They could reach 208mph and dash from rest to 100km/h in 3.95 seconds, while the corresponding figures for the 1,530kg rear-wheel drive SV were 201mph and 3.85 seconds. Sant'Agata had just given the Diablo a new lease of life.

The Diablo GT's front bonnet featured a strikingly protuberant trident shield.

The Diablo GT's convex bonnet shield and its twin roof-mounted air-scoops disrupt the supercar's otherwise clean profile.

THE DIABLO GT

Sant'Agata launched the Diablo GT on the opening day of the 1999 Geneva Motor Show with the extravagant claim that it was 'the world's fastest production car'. The GT moniker – which stood for Gran Turismo – was a complete misnomer, as the Diablo GT was a quasi-track car developed for speed, acceleration, and cornering. It was not a cossetting Grand Touring car in the classic mould, and it was singularly incapable of pampering its occupants over trans-continental-length journeys.

The series production, road-legal, Diablo GT could trace its ancestry back to three earlier racing Diablos – the SV-R, the GT1, and the GT2. The Diablo GT would follow in the tyre treads of these track-only cars with a stripped-out cockpit to save weight, and by adopting bodywork modifications that reduced aerodynamic drag and increased downforce, thereby increasing top speed, acceleration, and cornering speeds. The Diablo GT's performance was further enhanced by modifying its powertrain and braking systems for more power, torque, and retardation. And by carefully fettling the chassis, suspension, and suspension geometry, Sant'Agata gifted the GT better handling and roadholding than any of its predecessors, especially at the outer boundaries of its performance envelope.

The GT saved weight by reverting to rear-wheel drive, by utilising composites, carbon fibre, and other lightweight materials, by discarding all unnecessary in-cockpit luxuries, by using Alcantara instead of leather, and by employing featherweight alloy foot pedals.

Learning from the prototype Diablo GT2 race car, Sant'Agata's engineers increased the GT's Bizzarrini V12's displacement to 5,992cc (87mm bore and 84mm stroke), and designed a new lightweight crankshaft, a new intake manifold, new intake and exhaust valves, a modified exhaust system to reduce back pressure, and a new engine management system.

Aesthetic and aerodynamic highlights included a huge, shield-like protrusion that dominated the GT's front bonnet, the top end of which was an outlet vent for air scooped up by intakes low in the car's nose. Channelling this ambient temperature air past an oil cooler and along this aerodynamically computed route not only helped drivetrain cooling, but also created a venturi effect that produced genuine front-axle downforce. Four further styling features gave the GT a very purposeful and aggressive look: firstly, a redesigned front splitter-cum-bumper-cum-air dam incorporating three huge air intakes; secondly, a single very prominent roof-mounted air scoop that flowed backwards to merge with the engine compartment bonnet; thirdly, black carbon fibre sill extensions that ran from the front wheel arches to the rear arches, thereby increasing the car's effective undertray area and so increasing downforce; and finally a huge, black, carbon fibre rear wing with downturned end-plates and an adjustable central tab.

The GT's suspension geometry was completely revised with stiffer springs but softer dampers (in bounce), and the front and rear tracks were increased by 110mm and 30mm respectively. The size of the front Pirelli P-Zeros was increased to 245/35/ZR 18, while those at the rear remained unchanged at 335/30/ZR 18. The customer could choose between one of three gear sets depending on what they prioritised most – top speed, acceleration, or a compromise between the two. The GT was not a horse hair-shirt type of car – it boasted ABS brakes, air conditioning, power-assisted steering, and electronically controlled shock absorbers.

In contrast, Sant'Agata also built thirty Diablo GTR cars purely for the Lamborghini Super Trophy Championship. These rear-wheel drive GTR cars ran minimally modified V12 engines but were totally stripped of all luxuries and creature comforts, and were completely track-focussed with integrated pneumatic air jacks, roll cages, and X-door bars.

This Diablo GT's cockpit is swathed in lightweight Alcantara.

The Diablo GT was the fastest, lightest, most advanced, and most expensive series production car that Lamborghini had produced to date. All cars were left-hand drive, and Sant'Agata only homologated them for the European market – although a few found their way to the United States though grey market channels. The 1,490kg GT's 5,992cc engine developed 575bhp at 7,300rpm and 630Nm torque at 5,500rpm, and could power from rest to 100km/h in 3.7 seconds en route to a top speed of 211mph. The Diablo GT was the last Lamborghini conceived and developed in the pre-Audi era.

THE DIABLO 6.0 VT

The Diablo 6.0 VT was Ferdinand Piech's first opportunity to put his stamp on Lamborghini. Piech, the formidable and fearsome Chairman of the VW-Audi conglomerate, was not only the grandson of Ferdinand Porsche (the driving force behind the Volkswagen Beetle and the founder of the eponymously named sports car company), but was also a renowned engineer in his own right – he had designed the all-conquering Porsche 906 and 917 racing cars. Piech's business credentials were equally impressive – he had saved the Volkswagen group from bankruptcy, after which he had acquired Bentley, Bugatti, and Lamborghini for the conglomerate. Piech knew his mind, and he also knew exactly what he wanted from Sant'Agata. And its offerings as of September 1998, which is when Audi took over Lamborghini on behalf of the VW Group, were not to Piech's taste.

Piech would ideally have liked to launch a brand new model immediately upon acquiring Lamborghini, but designing a worthy successor to the Miura, Countach, and Diablo was not the work of a moment, especially as rival supercar manufacturers had made huge strides during Lamborghini's cash-starved period when it was variously owned by MegaTech, V Power Corporation, and MyCom Bhd. Piech was experienced and pragmatic enough to recognise that he needed to be patient while Sant'Agata's engineers and designers came up with a credible new flagship, but he was also impatient enough to demand that Lamborghini design and launch a comprehensively modernised and improved Diablo variant in the meantime. The Diablo 6.0 VT was the result.

For what would be a temporary replacement car, the Diablo 6.0 VT garnered universal and almost unadulterated praise. Piech had appointed Luc Donckerwolke as Lamborghini's chief designer, and the 6.0 VT that Donckerwolke presented at the Detroit Auto Show in January 2000 was a clean, arrow-like car, bereft of tacked-on ducts, vents, wings, air scoops and other ungainly protuberances. The car's exterior panels were almost all made of carbon fibre, except for its aluminium guillotine doors and its steel roof. A simple, more rounded, and less cluttered front bumper with two large air intakes for front brake cooling sat immediately below new and distinctive clear-lens headlights and indicator lights. The front bonnet was similarly uncluttered without any scoops or ducts. The 6.0 VT's rear was very tidy too, with a massive, single-piece, upward-slanting, rear bumper-cum-diffuser, two exhaust pipes, four tail lights, and a flat engine compartment bonnet with an octet of air vents. The side sill air intakes in front of the rear wheels were tidied up, while the air scoops above the rear wheels were smoothed out. A rear wing was a functionally unnecessary appendage and so was not a standard fitment, but it was available on the options list for those customers intent on messing up Donckerwolke's elegant design. The 6.0 VT's bodyshell looked thoughtfully designed, harmonious, and aerodynamic.

Carbon fibre was the dominant theme for the 6.0 VT's cockpit. A thorough interior redesign brought in a new crescent-shaped dashboard

The formidable Ferdinand Piech's triumph – the equally formidable Diablo 6.0 VT.

Seen here in Rosso Vik hue, Piech demanded that the Diablo 6.0 VT meet Audi-VW's exacting standards. By all accounts his wish was fulfilled.

that screamed out Audi. Which was no bad thing bearing in mind that Audi at this point in time was renowned for its beautifully crafted switches, dials, and gauges. Seven alloy-rimmed instrument gauges were set into a blue-black carbon fibre dash panel, while a shining, gear lever topped by an equally gleaming spherical aluminium gear knob sat in a circular, silver-coloured, open-slot gear gate, which itself was set into a polished, black, carbon fibre central console. Ergonomics was another Audi strongpoint, and all the dashboard gauges could now be read easily, while all the switches, buttons, and knobs could now be reached quickly and effortlessly. The driver and passenger airbags, the windscreen wiper system, the automatic climate control system, and the seats were all redesigned – an indication of just how seriously Piech, Donckerwolke, and Sant'Agata were taking their customers' safety and comfort.

Lamborghini's latest flagship car was powered by a highly modified, 5,922cc version of the classic 60 degree Bizzarrini V12 engine – the key developments being a lightened crankshaft and titanium conrods. The Lamborghini Iniezione Elettronica (LIE) engine management system now gained a 32-bit microprocessor in place of the previous model's 16-bit item, and this allowed it to better control the engine's sequential multipoint fuel injection system, the static electronic ignition system, and Lamborghini's new Exhaust Noise Control System (ENCS). The last-mentioned system allowed the 6.0 VT to scream and bellow in typical Raging Bull tradition while still conforming to current and forthcoming noise legislation.

Highlights of the 6.0 VT included an upgraded, viscous-coupled, four-wheel drive system; additional carbon fibre bracing for the spaceframe chassis; revised suspension geometry complimented with electronically controlled shock absorbers; widened front and rear tracks (by 60mm and 30mm respectively); and new, five-hole, monoblock, lightweight, magnesium alloy wheels measuring 8.5 × 18in and 13 × 18in, draped with 235/35 ZR18 and 335/30 ZR18 Pirelli P-Zero tyres at the front and rear respectively.

With 550bhp at 7,100rpm and 620Nm at 5,500rpm, the 1,625kg 6.0 VT could accelerate from rest to 100km/h in 3.95 seconds and reach a top speed of 205mph. The motoring press and customers alike were delighted with Lamborghini's latest flagship. But most significantly, the notoriously hard to please Ferdinand Piech, who had taken an active part in the 6.0 VT's development programme, including as an occasional test driver, was reported to be satisfied with his latest acquisition's newest offering.

CHAPTER 5

THE DEVELOPMENT OF THE MURCIELAGO AND THE EVOLUTION OF ITS MODEL VARIANTS

The Lamborghini Murcielago came into existence, just like the Countach and the Diablo before it, simply because its predecessor model had reached the natural end of its evolutionary cycle. Further development was unpragmatic simply because the changes needed to keep up with imminent legislation, customer expectations, and contemporary and forthcoming rival supercars were massive. In addition to Lamborghini's traditional competitors – Ferrari, Porsche and Aston Martin – Sant'Agata was now facing threats from unexpected, unheard of, and long-forgotten, quarters. Over the Murcielago's lifespan, Lamborghini would also have to battle with Pagani, Koenigsegg, McLaren, and Bugatti.

Furthermore, Ferdinand Piech, the ferociously ambitious Volkswagen Group Chairman, was fiercely protective of his automotive conglomerate's hard-won reputation for state-of-the-art engineering, mechanical reliability, excellent active and passive safety, beautiful interior and exterior aesthetics, and class-leading ergonomics. The Diablo 6.0 VT that Piech had so carefully watched over had gone some way in addressing the Diablo's inherent weaknesses, but it did not entirely meet Piech's exacting standards. Sant'Agata's engineers recognised that a step-change in product development was now urgently needed. Excepting Bugatti, Lamborghini was the apex manufacturer within the Volkswagen Group, and the Diablo's successor would sit at the very pinnacle of Lamborghini's offerings. The Murcielago had to be better than excellent.

Luc Donckerwolke designed the Murcielago's bodyshell, while Maurizio Reggiani headed the engineering group that developed the Murcielago's drivetrain, suspension, and mechanical ancillaries.

Reggiani was actually first tasked with developing a Diablo successor way back in 1995 when Lamborghini was still owned by the Indonesian V Power Corporation and the Malaysian MyCom Bhd companies. Marcello Gandini, the designer of the Miura, Countach and Diablo, submitted a full-size styling model but Sant'Agata rejected the maestro's

The Murcielago VT 6.2 Coupe showcased elements of the Countach – particularly in its trapezoidal front bonnet and its voluptuous front and rear wings.

The Murcielago VT 6.2 Coupe featured a Venetian-blind style engine compartment bonnet a la the Miura.

proposal in favour of a design from Zagato. At that time CAD-CAM (computer-aided design – computer-aided manufacturing) was the design buzzword, and Zagato was the only Italian design studio with the computer-controlled milling machinery capable of taking design proposals from computer screens and converting them directly into physical models. This Zagato design, which would in time be called the Lamborghini Canto, was striking but ungainly, although many might say that it would be more exact to call it strikingly ungainly. It had two terrifyingly gigantic air scoops high above its rear flanks like a fighter jet's intakes, and the words cohesion and elegance were strangers to the Canto lexicon. Despite all this, prototype Cantos were racing around the Nardo test circuit by 1997, and in the early months of VW-Audi's ownership of Lamborghini the conglomerate poured money into the Canto's development. Lamborghini were so convinced that the Canto would make it into series production that they announced that it would be launched at the 1999 Geneva Motor Show. But when faced with the Canto, Ferdinand Piech detested it at first sight, saying that it lacked fire, beauty, and drama, and that it did not stand true to Lamborghini's heritage. Redesigns, facelifts, and mechanical modifications and alterations did nothing to gain Piech's favour and the Canto was canned without ever being officially presented to the public.

At Piech's behest, Luc Donckerwolke had been parachuted into Sant'Agata as Lamborghini's new Head of Design in 1998. Donckerwolke had previously made his reputation working on the Audi R8 Le Mans Racer and the Audi A2 concept and production cars. Immediately after finishing the styling of the Diablo 6.0 VT, Donckerwolke started designing the Murcielago. Donckerwolke was honest and clever enough to acknowledge that he closely examined and drew inspiration from the Miura, Countach, and Diablo when designing the Murcielago. The Venetian blind engine compartment bonnet was pure Miura, while the trapezoidal theme that populates the Countach was mirrored throughout the Murcielago. The Countach's and Diablo's extreme wedge shape and their guillotine doors were by now signature Lamborghini design features and Donckerwolke sensibly retained both. There is never any shame in humbly acknowledging a maestro's works, and then gently but concretely evolving them so that they remain the fittest and thus deserve to survive.

Concurrently, Reggiani decided that the Murcielago would be a mechanical evolution of its previous two ancestors, the Diablo and the Countach. Reggiani had been employed by Paolo Stanzani, the father of the Countach, to work alongside him on the Bugatti EB 110 project, and Reggiani worshipped Stanzani and recognised just how revolutionary and groundbreakingly important Stanzani's South-North engine-gearbox orientation was to Lamborghini's heritage and to producing a competitive world-class supercar. Reggiani also understood that the Bizzarrini V12 engine was critical to Lamborghini's credibility and history, and appreciated the engineering and dynamic benefits of a rear mid-engine layout and a longitudinal orientation of the massive Bizzarrini V12 engine. All these features were therefore retained for the Murcielago.

THE MURCIELAGO VT 6.2 COUPE

The Lamborghini Murcielago VT 6.2 Coupe was first presented to a select group of potential customers and motoring journalists at about 10 p.m. on Friday, 7 September 2001. This launch took place on a cold and very windy evening in the foothills of a steaming and lava-spouting Mount Etna volcano in Sicily, and was set to the Biblical theme of the Seven Deadly Sins. On the Saturday an extravagant presentation and gala dinner took place at the Sant'Agata factory for a rather less select audience. Your authors were invited to this second event courtesy of having acquired, on 26 May 2001, temporary custodianship of Countach 88 ½ QV Chassis JLA12399 from H.R. Owen, at that time the sole UK Lamborghini Main Dealer. As excited as we were about being in Sant'Agata, and of having almost unfettered access to every corner of the factory, and of seeing the Murcielago and multiple examples of almost all its predecessors, the knock-out highlight of the trip for both of us was meeting and spending time with Paolo Stanzani, the father of the Countach. We would have to wait a further twenty-two years before meeting the Countach's mother – Marcello Gandini. The fingerprints of both these men, long-parted from Lamborghini, could be found all over the brand-new Murcielago.

Carrying the internal factory designation L147, the Murcielago was the first Lamborghini designed in its entirety by Luc Donckerwolke, and the first all-new Lamborghini produced by Sant'Agata while under the ownership of the VW-Audi Group.

Throughout the gala dinner rumours were flying around the crisp white linen-draped tables that the Murcielago featured a revolutionary and startling innovation, but everyone in the know had been sworn to secrecy and they all remained tight-lipped. Then a stunning 'Verde Ithaca' (fluorescent green) Murcielago was driven up an inclined stage set right among all the dining tables, and the car performed its party trick. Yes, everyone expected the car to be wedge-shaped and have guillotine doors, and we all knew that retractable headlights had essentially been outlawed by legislation, so there were no surprises here. But what no one had expected was the Murcielago's huge electronically controlled flank air intakes that could be raised and lowered at will. This was pure theatre and earned Lamborghini's latest flagship instant credibility from the assembled diners, and would go on to become one of the Murcielago's signature features, recognised and loved by enthusiasts worldwide.

BODYSHELL AND INTERIOR DESIGN

Luc Donckerwolke has stated that he wanted to take a purist approach when designing the Murcielago, but that his brainchild had to have within it the DNA of the Miura, Countach, and Diablo. To this end he retained the Countach's and Diablo's extreme wedge shape including their sharp noses and steeply raked windscreens, their sharply cut-off Kammback tails, their large-diameter exhaust pipes, and some air inlets and vents – although he took care to carefully and elegantly integrate these into the bodyshell both for aerodynamic and aesthetic reasons. And, of course, their guillotine doors. The Murcielago's doors opened a full 5 degrees more than the Diablo's and its sills were 25mm lower to boot, which allowed for easier and more graceful ingress and egress.

Much of the L147's interior was styled by Ralph Kluge, who had worked on the Audi Rosemeyer concept car, and his surprisingly sombre and conventional cockpit design stood in pleasingly sharp contrast to Donckerwolke's dramatically styled exterior. Seats with arrow-shaped headrests, a central console with an array of buttons and a silver-coloured gear lever topped with a spherical gear knob, and ahead of this a vertical face with audio and heating and ventilation controls, were amongst the interior highlights. The passenger enjoyed the luxury of a grab handle attached to the central console, an almost exact facsimile of that found in the Miura, while the driver could hang on to a small, perfectly round (eschewing the contemporary trend for flat-bottomed steering wheels) MOMO steering wheel during high-G cornering. Later

The Murcielago VT 6.2 Coupe was as curvaceous as it was sleek. Its multi-coloured taillights were one of the defining features of the early Murcielagos.

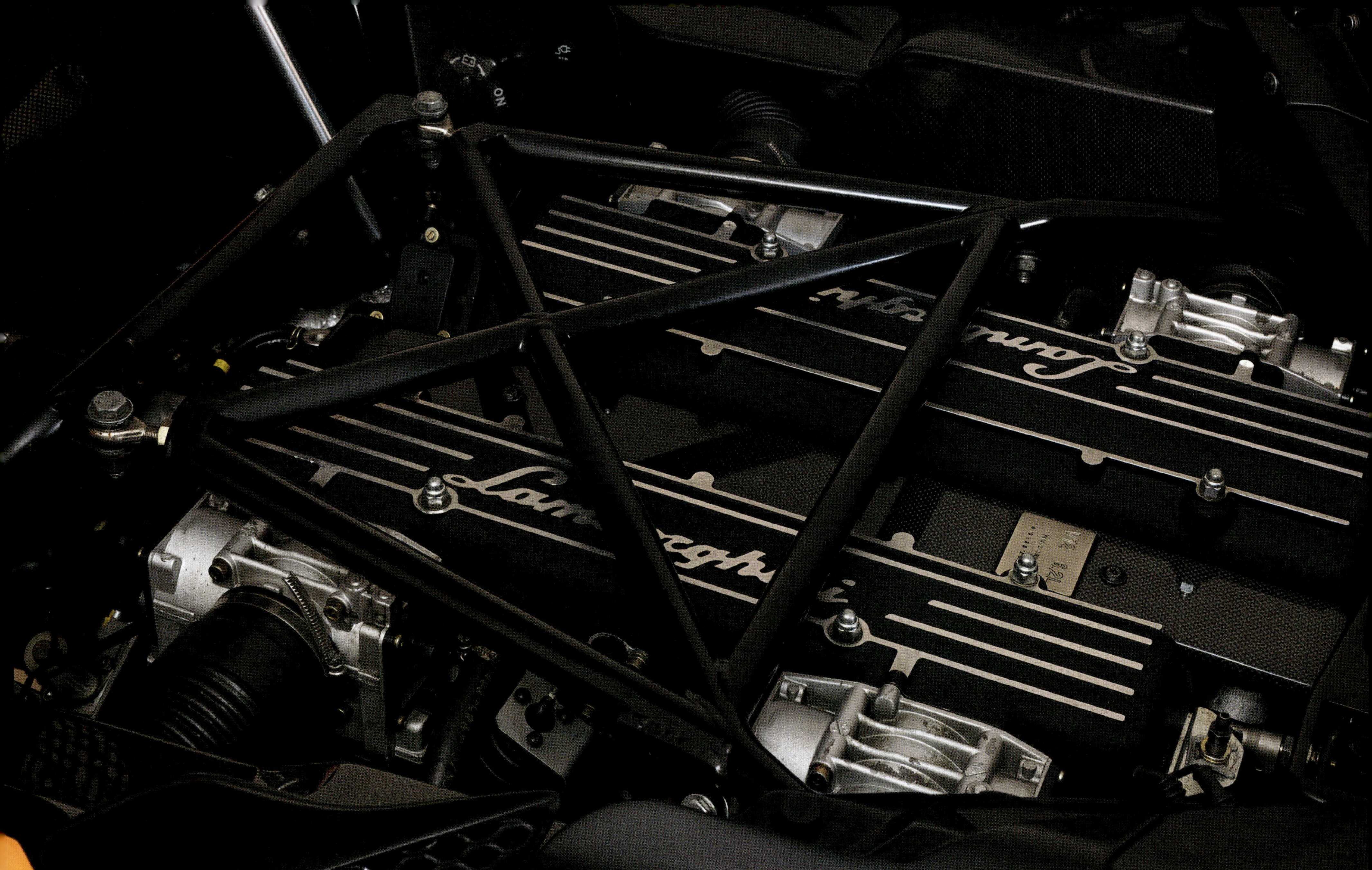

The powertrain of a Murcielago VT 6.2.

E-gear cars had two small, steering column-mounted alloy paddles behind the steering wheel. The angled dash binnacle housed two primary dials for the speedometer and the rev counter, four secondary dials for the oil pressure, the oil and coolant temperatures, and the fuel level, and no fewer than thirty-four multicoloured dashboard warning and indicator lights that could put a rainbow to shame. Leather abounded everywhere. Audi's influence could be seen throughout the cabin – in the ergonomics, in the quality of the cockpit furniture, and in the standard of the fit and finish.

SPACEFRAME CHASSIS AND SUPERLEGGERA CONSTRUCTION

The L147 was the last flagship Lamborghini to use the classic Superleggera chassis set-up. A key design objective for the Murcielago was that its spaceframe chassis should have exceptional torsional rigidity, as this would have a direct impact on the car's roadholding and handling through better suspension control, while also reducing noise and vibration transmission into the cockpit. To this end the spaceframe was constructed from high-tensile-strength, square-section steel tubes and married to a structural steel roof and a rigid carbon fibre floor pan. Carbon fibre and composite reinforcing struts were used wherever necessary, and were attached to the spaceframe with either steel rivets or adhesives. By utilising all these technically advanced materials and techniques, the Murcielago's chassis realised a highly commendable torsional rigidity in excess of 20,000Nm/degree. The L147's bodyshell was largely made from carbon fibre, except for the external skins of the guillotine doors, which were fabricated from aluminium, and the roof panel, which was made of steel.

ENGINE

The Murcielago 6.2 VT Coupe was powered by an evolution of Lamborghini's stalwart Bizzarrini engine. Developed by Massimo Ceccarani and carrying the internal factory code name L535, this all-aluminium alloy, 60-degree, V12 powerplant had a bore of 87mm and a stroke of 86.6mm for a total displacement of 6,192cc. The original plan was for the Murcielago to have a displacement of 5,992cc, exactly the same as the last of the Diablos, but this was changed for two reasons. Despite its extensive use of lightweight composite materials, the Murcielago weighed 25kg more than the Diablo 6.0 VT, and it was essential that the successor car had power-to-weight and torque-to-weight ratios that at least matched, or preferably exceeded, those of its predecessor. Secondly, Ceccarini and Sant'Agata's sales and marketing division wanted the Murcielago to be powered by an obviously different powerplant to the Diablo, and what better way to announce this change than by having a 6.2-litre engine rather than a 6.0-litre powerplant.

The single most important and remarkable change that the L535 engine featured over all its predecessors was its adoption of dry-sump lubrication. The original prototype engine that Giotto Bizzarrini produced for Ferruccio Lamborghini way back in 1963 had been dry sumped, but Giampaolo Dallara and Paolo Stanzani had been forced to remove this exotic feature in the process of taming Bizzarrini's quasi-racing engine and making it fit for use in a road-only Gran Turismo car. So the L535 was a throwback to Lamborghini's earliest days. Dry-sump lubrication was an important modification because it allowed the Murcielago's engine to be positioned a whole 50mm lower within the chassis, thereby lowering the car's centre of gravity and so improving the car's stability, roadholding, and handling. Furthermore, placing the engine lower down allowed Luc Donckerwolke more space and more leeway to optimise the car's aerodynamic profile by altering its rear bodywork.

A key priority for the L535 engine was that it could meet the ever more stringent emission standards that were being introduced in all of Lamborghini's key markets, particularly the United States, Northern Europe, and Japan. New innovations that allowed the L535 to conform with these regulatory requirements included an oil scavenger pump with double the oil pressure pump's capacity, an aerator to remove air and foam from the lubricant oil, an electronically controlled variable geometry air intake system (VIS), and an electronically controlled variable valve timing system (VVT) that operated on both the intake and outlet valves. The quad camshaft, four valves per cylinder, L535 engine operated at a high compression ratio of 10.7:1 and utilised static ignition with one coil for every sparkplug. By exploiting these developments the L535 produced high torque over a relatively wide rev range – Lamborghini claimed that the Murcielago produced more torque at just 2,000rpm than the peak torque achieved by the majority of its contemporary supercar competitors. Good fuel consumption, low emissions, and superb drivability (helped here by the Murcielago's new and excellent 'drive-by-wire' electronic throttle control system) were the other prizes gained by adopting and exploiting these technological innovations. Sant'Agata released the following fuel consumption figures for the Murcielago: City – 32.6 litres/100km; Highway – 15.11 litres/100km; Combined 21.5 litres/100km.

This L535 engine with a specific output of 92bhp per litre produced 571bhp at 7,500rpm and 650Nm torque at 5,400rpm, giving the 1,650kg Murcielago an approximate power to weight ratio of 343bhp per tonne. This was enough to propel the Murcielago 6.2 VT coupe to a top speed of 205mph, and to power it from rest to 100km/h in about 3.8 seconds.

GEARBOX

The Murcielago's brand new gearbox represented another first – the first six-speed gearbox that Sant'Agata had ever put into a series production car. This gearbox was made by CIMA and was used in modified form in the Gumpert Apollo and Pagani Zonda hypercars. The six forward gears were set in a conventional double-H pattern, while engaging reverse gear required the driver to first press the gear knob downwards and then to the left and forwards. To increase the transmission system's rigidity and stability this gearbox's primary and secondary shafts were each mounted on three (rather than on two as in the Diablo) bearings, while to improve gear-change feel and reliability new double- and triple-cone synchronisers were introduced together with modified control linkages. Depressing the Murcielago's clutch still required a hefty downward push, but this was much less of a Herculean affair than in the Diablo, let alone the Countach. This was achieved, in part at least, by repositioning the clutch slave cylinder in line with the clutch release bearing axis.

E-GEAR

In 2004 Sant'Agata started offering the E-Gear system – a robotised manual transmission system otherwise known as an automated manual transmission (AMT) system – as an expensive optional extra on the Murcielago. Based on the Magneti Marelli system used in contemporary Formula One racing cars, and pioneered for road use by the Ferrari 355 F1 in July 1997, this electro-hydraulic arrangement dispensed with a clutch pedal and a gear lever, and instead allowed the driver to choose higher or lower gears by operating steering column-mounted alloy

The cockpit of a Murcielago VT 6.2. Note the steering column-mounted paddles for the E-Gear transmission system.

A Murcielago VT 6.2 Roadster with black Hercules wheels.

paddles. Pulling on the left paddle instructed the transmission system to drop down a gear, while pulling on the right paddle instructed it to select the next highest gear, and pulling on both paddles simultaneously selected neutral. Reverse gear was selected by pressing a button on the lower dashboard. An electronic control unit controlled clutch disengagement, gear engagement, and clutch re-engagement through specific, very high-pressure hydraulic actuators. With the E-Gear system fingertip pressure replaced forearm and thigh heft. Other advantages of the E-Gear system included the safety aspect of being able to keep both hands on the steering wheel during gear changes, faster gear change speeds than what Formula One racing drivers could achieve manually, and the possibility of sequential and very, very rapid down changes on the approach to a corner. Disadvantages included low-speed jerkiness and the supposed loss of driver involvement. Your authors would say that the latter disadvantage is a dubious and over-hyped objection in any very high-powered and heavy supercar – but it is eagerly touted by second-hand car dealers selling manual transmission Murcielagos. The antiquated single-clutch robotised manual system in the Murcielago

All shut down and as sleek as can be.

Guillotine doors up.

is nowhere as fast or as refined as the latest dual-clutch transmission systems, but it has its own honest charm, especially in its mechanical feel and sound on gear engagement and disengagement. The E-gear system was a real advance in transmission technology, as evidenced by the fact that almost every reputable supercar manufacturer now only offers clutchless transmission systems.

The E-Gear system in the Murcielago allows the driver to choose between a number of different driving modes. In fully automatic mode an electronic control unit decides on the optimal gear and then sends an electronic signal to the gear actuators instructing them to carry out this decision. Alternatively the driver can take over and decide which gear is best and can then manually select this gear using the steering column-mounted alloy paddles. Within this second mode the driver can select between three different driving programmes – Normal, Sport, and Slippery Road. In the Normal setting the car automatically engages the next highest gear when the car's engine speed reaches 7,450rpm. In the Sport setting the there is no automatic up-shift and gear changes are executed more quickly. And in the Slippery Road setting the system uses softer pick-up points for first, second, and reverse gears, automatically engages the next highest gear once the engine speed reaches 3,200rpm, and sets a lower threshold before the traction control system intervenes.

FOUR-WHEEL DRIVE VISCO-TRACTION SYSTEM

The Murcielago used a refinement of the all-wheel drive Visco-Traction (VT) system found in the Diablo. A central viscous coupling working in tandem with self-blocking differentials allows a 25% limited slip at the front and a 45% limited slip at the rear, thus minimising the possibility of a loss of traction and a secondary loss of control. The VT system does this by directing any excess torque from the primary axle (the rear axle) to the secondary axle (the front axle) whenever the rear tyres start to spin. In addition to this the Murcielago featured a completely separate electronically managed traction control system that works by reducing engine torque through electronic intervention on the fuel-injection and ignition systems, and also by electronically intervening on the throttle through the drive-by-wire system. Expert or over-confident drivers could manually switch off this electronic traction control system by pressing a button on the central console.

The 6.2 VT coupe had new multi-piece, five-hole, 18inch, aluminium alloy wheels measuring 8.5 × 18inches at the front and 13 × 18inches at the rear. These were wrapped with Pirelli P Zero Rosso front and rear tyres measuring 245/35 ZR 18 and 335/30 ZR 18 respectively. Lamborghini recommended that for normal use these tyres should be inflated to 2.6 bar for the fronts and 2.5 bar for the rears, except for speeds above 290km/h when all four tyres should be inflated to 3.2 bar.

SUSPENSION

The Murcielago's not inconsiderable 1,650kg was distributed in an almost ideal pattern: 48% front and 52% rear. This weight was carried on a classic, independent, front and rear double wishbone suspension set-up. Six state-of-the-art shock absorbers, each with electronic damping control and shrouded by a coaxial spring, were used – one for each of the two front corners, and two for each of the two rear corners. Each of these damper-spring units was attached to the chassis through Flanbloc bushes. Front 'Antidive' and rear 'Antisquat' technology acting in concert with front and rear anti-roll bars ensured that the car remained as level as possible during hard acceleration, heavy braking, and high G-force cornering. An essential optional extra for protecting and preserving the Murcielago's expensive and vulnerable carbon fibre front spoiler-cum-bumper was the electrohydraulic front lifting system. When activated by pressing a button on the central console, this system raises the front of the car by 40mm. This front lift system could only be summoned into action when the engine is running and when the car is travelling at no more than 70km/h (or 30km/h for Japanese market cars). Once raised the car could be brought back to its normal ride height by pressing this same button again. Owners were strongly advised never to leave their cars for extended periods with the front suspension elevated, as doing so places undue stress on the shock absorber seals.

BRAKING SYSTEM

The Murcielago VT 6.2 Coupe boasted two totally independent hydraulic braking systems, one for the front axle brakes and the other for the rear axle brakes. Safe braking from high speeds, or in slippery conditions, or on uneven surfaces, was optimised by employing a four-channel ABS system, a Dynamic Rear Brake Proportioning (DRBP) system, and by integrating the braking system with the car's traction control system. The TRW four-channel ABS system monitors and controls each wheel independently using hub flange-mounted electromagnetic wheel speed sensors. These sensors continuously send signals to a dedicated microprocessor that calculates the speed and acceleration of each wheel and thereby detects if a tyre is about to lose traction and spin. If any tendency to slip is detected, the microprocessor activates an electrohydraulic control unit that modulates the brake pressure in the appropriate brake line to brake that particular errant wheel and thus stop it from slipping. The DRBP system in the Murcielago was a further development of the system found in the Diablo, and it optimally distributes the available braking force between the front and rear tyres at all times, and in all conditions.

STEERING, SAFETY, AND VISIBILITY

The Murcielago had a rack and pinion steering system with hydraulic power assistance. The steering wheel twirls through about three turns from lock to lock, and the car's turning circle is about 12.5 metres. For safety reasons the steering column is collapsible, and a 60-litre airbag is mounted within the car's steering wheel boss. Passenger safety is catered for by a 120-litre airbag hidden in the dashboard. The driver is alerted to any malfunction in the airbag system by a red flashing light in the dash display. Lamborghini strongly emphasised that these airbags do not replace the three-point, cross-abdominal, safety belts with their in-built pyrotechnic belt tensioners, which should always be used. Bi-Xenon headlights dealt with low ambient light conditions, while the windscreen and headlights were kept clean and clear by a single-arm pantograph windscreen wiper and high-pressure headlight washer jets respectively. Electrically foldable wing mirrors and a non-reflecting cockpit rear-view mirror provided the driver with some information of the threats surrounding his Italian steed – rear and side visibility being long-standing Lamborghini weak points. The extended wing mirrors also give the driver a good indication as to whether the Murcielago can pass safely through a given gap – the rear of the car is considerably wider than its front, but when fully extended the mirrors are wider tip-to-tip than the rest of the Murcielago.

The Murcielago Roadster's engine compartment bonnet was rear-hinged, like the Miura's, and unlike its hard-top counterpart's.

One of the 'Specials' – a 40th Anniversary Murcielago.

The Murcielago LP 640 Coupe had asymmetrical sill air-intakes.

THE VARIABLE AIRFLOW COOLING SYSTEM (VACS)

The L147's party trick was its Variable Airflow Cooling System. Most modern petrol-driven internal combustion engines are relatively inefficient, with only about 25% of the inherent chemical energy of the petrol actually converted into the kinetic energy that propels the vehicle forwards. The remaining 75% of this energy is converted into heat energy (or left as uncombusted fuel), which can be very difficult to expel quickly and efficiently out of the tight confines of an engine bay. Sant'Agata's engineers had repeatedly come across this problem when designing and developing the Miura, Countach, and Diablo, so they had every reason to expect the Murcielago's massive 6.2-litre engine to present them with a similar challenge. Liquid cooling is far better than air cooling at dissipating an engine's heat, but even so this heat energy ultimately needs to be transferred to the atmosphere, and a 6.2-litre V12 engine therefore needs large volumes of constantly renewed ambient temperature air. This in turn demands large air intakes, but air scoops are inherently drag inducing, especially if they directly face the oncoming airstream. A lesser but still important consideration, especially for a supercar, is that large air intakes and vents are generally ugly and clutter the clean lines that customers expect.

Lamborghini neatly resolved the contradictory demands of aerodynamic efficiency and clean aesthetics with the challenge of getting adequate volumes of ambient temperature air into the engine compartment by designing a system in which the air intakes stay hidden and flush with the car's sleek bodyshell right up to the moment when more cooling air is genuinely needed. Lamborghini had worked out that under normal usage in most of its markets the Murcielago only needs large-aperture air intakes for 15% of the time. For the remaining 85% of the time large air intakes would simply be an unnecessary aerodynamic and aesthetic burden. But in the Middle East and the tropics, and in stop-start city traffic jams, or on race tracks, Sant'Agata had to bite the bullet and somehow provide much larger volumes of cool air. The ideal solution was therefore to have variable-aperture air intakes that could be large when high cooling is required, and small at all other times. When designing the Murcielago's Variable Airflow Cooling System, Massimo Ceccarini looked to, and got inspiration from, a previous Lamborghini design concept car that never made it into production – the mid-to-late 1980s Countach L150 Prototipo, in which Giulio Alfieri replaced the Countach's signature flank NACA ducts with movable side air intake flaps. The Murcielago's VACS is made up of sensors that read the engine coolant and ambient air temperatures and send the appropriate signals to a dedicated ECU, which in turn controls direct current electrical motors that either open or close the movable air intakes sitting astride each of the Murcielago's rear flanks. Under normal circumstances these air intakes stay lowered and sit flush with the body, but can rise to a 20 degree position. When raised to this 20 degree position the air intake surface area increases by 80%, so providing a huge amount more of ambient temperature air for cooling the drivetrain. These movable lateral air intakes only have binary positioning – either fully open or fully closed. As well as the automatic raising and lowering of these intakes as dictated by the ECU, Lamborghini recognised that some owners might favour form over function and prefer the aesthetics of having these flank air intakes permanently open. Sant'Agata therefore thoughtfully provided the driver with a push-button switch on central console allowing him or her to raise or lower these flank intakes at will. When viewed head on the raised VACS air intakes look like the spread-out wings of a bat, or like Batman in his classic stance with his cape held wide open. By happenchance the word Murcielago is Spanish for bat, and in two Warner Brothers films, *Batman Begins* and *The Dark Knight*, the Murcielago features as Batman's steed of choice when sans mask and cape, and when not driving his rocket-firing Batmobile.

A Lamborghini Murcielago Roadster LP 640 in vibrant Verde Ithaca. Note the facelifted sill air-intake and the new silver Hermera wheels.

The Murcielago LP 670-4 SV was a purist's car aimed squarely at the driving enthusiast. It was the last Lamborghini to be powered by Sant' Agata's classic Bizzarrini V12 engine.

ACTIVE REAR SPOILER

Lamborghini's latest flagship featured an active, three-position, electronically controlled rear spoiler. To minimise aerodynamic drag and yet keep the car stable at high speed, this rear spoiler sits flush with the bodywork until the car reaches 130km/h, at which point the spoiler raises itself automatically to a position of 50 degrees from the horizontal. Once the car reaches 220km/h the spoiler rises further to its final resting position of 70 degrees from the horizontal.

THE ELECTRONIC MANAGEMENT SYSTEMS

The Murcielago needed a much more extensive and sophisticated diagnostic, logging, and control electronic management system than the Diablo or the Countach. This was largely because the Murcielago had to comply with much stricter and wider-ranging contemporary legislation – mainly regulations pertaining to noxious emissions, and driver, passenger and pedestrian safety. A lesser driving force for the Murcielago needing such a sophisticated electronic management system was so that it could meet customer expectations relating to the car's performance, reliability, and comfort.

Four principal electronic control units, each using a 32 bit/20Mhz microprocessor, controlled a vast variety of functions. There were two Lamborghini LIE management units, each controlling one of the two engine banks. There was also one Lamborghini GFA management unit and one Lamborghini Power Motor management unit, and these controlled what Lamborghini termed 'body' functions. Amongst the many functions that these four principal units controlled were the multipoint sequential fuel-injection system, the ignition management system, the variable valve timing of both the inlet and outlet valves, the variable geometry intake management system, the detonation management system, the electronic throttle system, the variable airflow cooling system, the rear spoiler deployment and control system, the three-way catalytic convertor and oxygen sensor and emissions management system, and the OBD II on board diagnostic system.

Ferdinand Piech had two prime objectives for Lamborghini when he bought this Emilia Romagna-based supercar manufacturer. Firstly, he wanted Lamborghini to be a quality manufacturer whose products could sit proudly alongside the best of VW's other automotive offerings. Piech ensured that this happened by imposing upon Lamborghini the sort of quality control regime that Sant'Agata could only have dreamt of previously. Giving substance to this assertion is the fact that Piech insisted that twelve test cars be built on Sant'Agata's Linea Uno (Production Line 1) before any series production cars destined for paying customers were built. Test cars were shipped to the United States for high temperature and low humidity testing in the desert, while other test cars were sent to Nardo and the Nürburgring for high-speed evaluation. No longer would early adopters act as unpaid test drivers for Lamborghini.

Piech's second prime objective was to see Lamborghini re-establish itself as a prime supercar manufacturer – and performance was the key to achieving this goal. Three prototype L535 engines were built and extensively bench-tested on Sant'Agata's dynamometers before Piech granted this engine his blessing. Each of these engines was found to reliably produce a peak power of 571bhp at 7,500rpm, and a peak torque of 650Nm at 5,400rpm, with the rev limiter kicking in at 7,800rpm. This equated to a power-to-weight ratio of 343bhp for the 1,650kg Murcielago, giving substance to Lamborghini's claim that the car could sprint from rest to 100km/h in 3.8 seconds before reaching a top speed 'in excess of 330km/h'. Piech was reported to be delighted with the Murcielago VT 6.2 Coupe, and Lamborghini was now well on its way to turning a consistent profit and to becoming a trophy subsidiary of the VW-Audi conglomerate.

THE MURCIELAGO VT 6.2 ROADSTER

Roadsters as a general rule sit at the very pinnacle of their model range. This is a long-established tradition – the most expensive, exquisite, and exclusive of Fiat's, Alfa Romeo's, Maserati's, Ferrari's, and Mercedes Benz's offerings from the 1920s through to the 1960s were open-top cars designed for the rich, the beautiful, the famous, and the well-connected. Lamborghini, that most avant-garde of supercar manufacturers, could not afford to disregard this niche but very lucrative market segment. The Murcielago VT 6.2 Roadster was launched in 2004.

Styling is all important in a roadster, particularly a supercar roadster, and amongst the VT 6.2 Roadster's contemporary rivals only the Pagani Zonda Roadster could hold a styling candle to it – albeit at almost double the price. Luc Donckerwolke said that he drew inspiration for the Murcielago Roadster from three quite disparate sources: the nuclear-capable B-2 Spirit stealth bomber; the 118 Wallypower superyacht, which can reach 60 knots while carrying six guests and six crew; and the Ciutat de les Arts I les Ciencies in Valencia, which is an architectural wonder housing an opera house, an arts centre, and a science museum among many other attractions.

Lamborghini first presented an open-top version of the Murcielago as a concept car at the 2003 Detroit Motor Show where the Murcielago Barchetta Concept instantly gained critical acclaim and immediately became the undisputed star of the exhibition. Barchetta is an Italian word meaning 'little boat', and was a term extensively applied to small open-top two-seater sportscars that eschewed luxury features and styling excess in favour of minimising kerb weight to maximise performance. The irony remains hard to dismiss.

The Murcielago VT 6.2 Roadster featured new and dramatic exterior styling, luxury interior upgrades, and state-of the-art safety innovations. It also suffered from a slightly higher kerb weight and ran softer suspension settings compared to its coupe sibling – all of which would hurt rather than enhance performance.

Donckerwolke has stated that he does not like open-top cars, not only because of their degraded torsional rigidity, but also because they lose a key styling feature – the roofline's clean sweeping arc that runs right across the centre of the car – which allows a designer latitude to express himself and his ideas. It is all the more wonderous then that the 6.2 Roadster ended up looking quite so sensational. Sitting just 42 inches tall with its soft top off, but 80.5 inches wide, the car's obvious beauty was suffused throughout by an air of aggression.

The Roadster's key exterior styling differences over its hard-top counterpart start with its windscreen, which is lower and more steeply raked than the coupe's. This expansive wraparound windscreen recalls the one-piece, curved windscreen of the Tipo 829 World Rally Championship-winning Lancia Stratos, which incidentally was designed by Marcello Gandini. Moving backwards, the windscreen merges seamlessly with the cut-down side window glass, sliding elegantly downwards to meet the bodywork just ahead of the of the retractable VACS air inlets. At the rear is the engine compartment bonnet, which is completely different to the coupe's, and which has regularly and repeatedly been called a work of art. This bonnet had to be totally redesigned as the coupe's bonnet could not accommodate the pop-up roll-over bar mechanism that is indispensable in this open-top car. The 6.2 Roadster's engine compartment bonnet pays tribute to the Miura in being rear-hinged, and to the Countach QV in mirroring that car's mille-feuille central hump. Another highlight of the Roadster was its new and gorgeous monoblock, five-spoke, Hercules wheels, which had more than a passing resemblance to the early Lamborghini Bravo wheels. But these Hercules wheels had function as well as form – they provided additional space for larger, upgraded brakes. The early Murcielago coupe's brakes had been condemned as being inadequate for very high-speed road use and track work, and Sant'Agata responded by extensively redesigning the Murcielago Roadster's brake system. New larger diameter (14.96 inches front, 14.00 inches rear) and thicker dual compound brake discs were married up to new eight-piston front callipers and a redesigned and more powerful brake booster. With these revisions the Roadster could decelerate from 125mph to standstill in just 130 yards, and brake fade was markedly reduced.

Losing the structural integrity of a fixed roof meant that the Roadster lost torsional rigidity, and Sant'Agata restored some of this by adding metal, composite, and carbon fibre bracing where necessary. Lamborghini strengthened the floor pan and the door sills at strategic points, and also mounted a lattice-like black-painted steel cage over the engine bay to provide additional rigidity to the Roadster's rear section. All this added an extra 15kg in kerb weight over the coupe, but for an additional £3,000 premium particularly rich customers could specify the engine cage in carbon fibre, thereby saving 5kg in kerb weight.

Losing the fixed roof also compromised driver and passenger safety in the event that the car overturned, and Sant'Agata designed Lamborghini's Active Roll-Over Protection System (AROPS) in response to this safety-critical issue. Fixed roll-over bars would have completely ruined the Roadster's clean and elegant lines, so a concealed, active system was designed. Roll-over sensors distributed throughout the car monitor the degree of vehicular tilt and send a signal to an electronic control unit once a set threshold is exceeded. This ECU then activates a mechanism that shoots up the two roll-over bars hidden behind the driver and passenger seats, while simultaneously triggering the seatbelt pretensioners. All this happens within milliseconds.

To reduce in-cockpit turbulence when driving al fresco, Lamborghini provided a windschott and a top spoiler. The windschott is a detachable glass pane with two locating pins that slot into a panel behind the seats. The top spoiler is a two-part item that attaches to the top of the windscreen through a rotating clasp mechanism and diverts the oncoming airstream away from the cockpit.

The Roadster's drivetrain was largely identical to its coupe sibling's, with the same engine, the same CIMA six-speed manual gearbox, and the same optional E-Gear robotised manual gearbox (which at launch commanded an £8,000 premium). But the Roadster had softer spring and damper rates, giving it a more cosseting ride. The single most notable feature of the Roadster's interior was its asymmetric leather upholstery – the driver's seat and door panel was clothed in perforated leather, while the equivalent passenger areas were draped in smooth leather.

The Roadster's Achilles heel was its detachable soft top. Fitting and removing this bulky item takes time, effort, expertise, and experience, and ideally requires two people, although one person can just about perform these tasks without damaging the car. The soft top is composed of four principal components. Firstly, the main fabric body that carries within it rubber gaskets, metal securing pins, Velcro strips, six press studs, two postero-lateral Tenax fasteners, and the glass rear window. The remaining three components are a right and left upright, and a removable central crossmember. All these parts have to be attached to the car or to each other in a pre-arranged sequence. Once assembled and fitted the soft top is superbly water tight and air tight, and no less cosseting than the coupe's fixed hard top. Lamborghini stated that 'the soft top was designed as

The F-22 fighter-jet inspired Lamborghini Reventon Coupe. Note its flat-plane body panels and the carbon fibre fins screwed into its wheel spokes.

emergency protection in case of rain', and warns that it is not guaranteed to stay attached to the car at speeds beyond 160km/h.

Your authors have designed their own bespoke tonneau in conjunction with Shamrock Trim in the United States. In wet and windy England, where torrential rain can alternate with bright sunshine at a moment's notice, the ten minutes it takes to erect the Roadster's soft top is nine minutes too long. Our bespoke tonneau is made up of a single layer of waterproof canvas with an integrated rain strip, two strong door mirror stem attachments, a couple of rear bodywork fasteners, and three metal bars. This tonneau does not require any modifications to the car at all, and can be put on and taken off in about 60 seconds. An additional benefit is that when packed up neatly this tonneau only takes up about 10% of the available front boot space, whereas the factory soft top takes up almost all of the luggage compartment.

The Murcielago VT 6.2 Roadster was, and remains, a styling triumph that matches and then surpasses its coupe sibling in almost every respect. With its 6.2-litre naturally aspirated V12 engine producing 571bhp at 7,500rpm and 650Nm torque at 5,400rpm, Sant'Agata claimed that the Roadster could accelerate from standstill to 100km/h in a scant 3.89 seconds on its way to a top speed of 199mph. It is no surprise that the startlingly elegant Murcielago 6.2 Roadster won starring roles in Hollywood – in the films *Transporter 2* and *Batman Begins*.

THE MURCIELAGO LP 640 COUPE

The Murcielago LP 640 Coupe made its debut at the Geneva Motor Show in March 2006. Its predecessor, the Murcielago VT 6.2, had enjoyed a five-year reign but this original was now facing a host of aggressive new challengers – not only from Ferrari with its 599, and Porsche with its Turbo and GT motorsport variants of the 911, but also from a new group of hyper-exotics. Hypercars like the Pagani Zonda, the Koenigsegg CCX, and the Bugatti Veyron, were pushing the performance envelope so far forwards that Sant'Agata had to improve upon the VT 6.2 simply to stay relevant in the supercar segment. And then, of course, there was a genuine in-house challenge from Lamborghini's lesser but still very potent and much more usable offering, the Gallardo. The Gallardo touted 500bhp and 510Nm of torque from its longitudinally mid-mounted V10, and was cheaper to buy and service, more fuel efficient, less polluting, and was much more manoeuvrable thanks to its considerably smaller footprint. Lamborghini was careful to make a clear distinction between its flagship car and its junior supercar by ensuring that the former was a much more bespoke product, and by reserving its signature guillotine doors only for the V12-engined Murcielago. To justify its price tag and to shore up its position at the top of the supercar food chain Sant'Agata now decided to restyle the Murcielago and improve its performance.

In restyling its flagship Lamborghini did not totally ignore function for form. Their designers knew that a larger-capacity, more powerful, engine was in the wings, and that this powertrain would need more cooling They also looked to improve upon the Murcielago VT 6.2's aerodynamics, knowing that this would have a direct and positive impact on performance, fuel efficiency, and emissions.

The LP 640 Coupe gained a new, more angular, front air dam-cum-bumper, with larger, five-sided air intakes. The front spoiler was now sharper and more prominent and featured a central indent that made for a far more aggressive head-on appearance. Moving backwards, the lower sill air intakes were dramatically redesigned, and most strikingly were asymmetrical. The left sill intake was larger and more prominent, and its posterior aspect was completely open, while the posterior aspect of the smaller and more discreet right sill intake was closed. There was a good reason behind this asymmetrical madness – the LP 640 had a single large oil cooler located on the left side of the car and the left sill air intake therefore had to provide a much greater volume of cool air. The rear of the LP 640 featured a new, single, very large, ellipsoidal, exhaust pipe, a new rear diffuser, new tail lights using LEDs rather than conventional incandescent bulbs, and new monocolour red tail light covers. The windscreen wiper system and the wing mirrors were redesigned to make them more aerodynamic. New aluminium alloy Hermera wheels measuring 8.5 × 18 and 13 × 18, and swathed in Pirelli P Zero Rosso tyres measuring 245/35 ZR 18 and 335/30 ZR 18 front and rear respectively, now gave the Murcielago a quite different appearance. Two very expensive optional extras were now available for this latest Murcielago – carbon ceramic brakes (£7,780 in 2006 money), and a transparent engine cover (£4,312).

The LP 640 boasted new, wider, and more cosseting seats, and the cabin leather covering the seats, door panels, transmission tunnel, and the roof panel featured a new lozenge-shaped quilting called Q-citura.

At the heart of the LP 640 was its modified drivetrain. It had long been thought that the Bizzarrini engine could not be stretched beyond 6.2 litres, but Sant'Agata's engineers worked their magic and produced a 6,496cc engine with a bore of 88mm and a stroke of 89mm. The primary design objectives for this facelifted engine was that it should produce more power and torque and be more flexible right across the rev range. With a completely new cylinder head design, a revised continuous variable valve timing system working on both the intake and exhaust valves, a new intake system, a new cooling system, a larger oil radiator, a new crankshaft, revised camshafts, and a modified exhaust system, this new engine could be coaxed to produce 631bhp at 8,000rpm and 660Nm torque at 6,000rpm. It also had a 300rpm higher rev limit than the VT 6.2's engine, redlining at 8,100rpm. This modified engine with a compression ratio of 11.0:1 and a specific output of 97bhp per litre gave the 1,665kg LP 640 coupe a power to weight ratio of 357bhp per tonne.

To cope with this increased power and torque the six-speed manual gearbox was modified with stronger internals, and the rear differential and axle shafts were replaced with stronger items. The LP 640's E-Gear transmission featured revised software to improve gear engagement and clutch life, and Lamborghini's launch control system, which Sant'Agata called Thrust Mode. The LP 640 also featured new springs, redesigned electronically controlled dampers, bigger front and rear anti-roll bars, and updated ECUs.

The LP 640 Coupe cost about 10% more than its predecessor, and for this added outlay customers got a mechanically and aesthetically facelifted supercar that could reach 211mph and sprint from standstill to 100km/h in just 3.4 seconds.

THE MURCIELAGO LP 640 ROADSTER

The stunning Murcielago VT 6.2 Roadster had been a huge commercial success and had also further enhanced Lamborghini's (and the VW Group's) already sky-high image and reputation, so it was always a given that an open-top version of the LP 640 would make an appearance at some stage. And the LP 640 Roadster did, just nine short months later, at the Los Angeles Motor Show in December 2006. It is telling that Sant'Agata decided to first show this Roadster, with its difficult to fit soft top, in sunny and dry California, rather than in wet and windy Europe. The LP 640 Roadster was an amalgam of the VT 6.2 Roadster and the LP 640 Coupe.

This new Roadster retained the old Roadster's shorter and more steeply raked windscreen, and its rear-hinged engine compartment bonnet, active pop-up roll-over bars, additional chassis bracing, and asymmetric cabin leather. From the LP 640 Coupe this new Roadster inherited the updated engine, transmission, drivetrain ancillaries, suspension, electronics and cabin improvements. The 640 Roadster enjoyed softer damper settings and spring rates than the coupe, giving its occupants a more comfortable ride over poor road surfaces. When the 631bhp, 660Nm torque, 1,665kg LP 640 Roadster made its entrance into the United Kingdom market it was priced at about £200,000 – at a time when the departing VT 6.2 Roadster and the then current LP 640 Coupe both cost about £190,000.

THE MURCIELAGO LP 670-4 SUPERVELOCE

Lamborghini aficionados throughout the world recognise that the Murcielago LP 670-4 Superveloce has two claims to fame. Firstly, it is a purist's car aimed squarely at the driving enthusiast, and hence has no unnecessary glitzy bits. It sits at the pinnacle of the Murcielago range, eclipsing in many respects even the hyper-expensive and ultra-showy Reventón. Secondly, it has historic provenance as the last Lamborghini powered by the classic V12 Bizzarrini engine – which had served Sant'Agata so well and so faithfully since 1963.

Speaking at the press conference following the 670-4 SV's launch at the Geneva Motor Show in March 2009, Stephan Winkelmann, Lamborghini's CEO, explained that Sant'Agata's latest offering was a strictly limited edition car and that only 350 units would be built. But the state of the world economy decided otherwise, and in the end only 186 of this most desirable of Murcielagos ever found first owners.

The 670-4 SV, aimed to be special through a combination of meticulous weight saving, increased power and torque, rarity, and exclusivity. The 1995 Diablo SV had jettisoned four-wheel for two-wheel drive to save weight, but Sant'Agata rejected this approach for the Murcielago 670-4 SV for a number of reasons. Firstly, Lamborghini's engineers felt that the 670-4 SV could only properly deploy all its power and torque if all four wheels were driven. Secondly, the added traction from the driven front wheels would help the car accelerate out of corners. Thirdly, all-wheel drive provides an extra layer of safety when driving a heavy, high-torque, wide-tyred vehicle in the damp, in heavy rain, and in icy conditions. And fourthly, Lamborghini's parent group had for decades been heavily invested into the quattro four-wheel drive system, and it would not do for the flagship car of one of VW-Audi's most high-profile sub-brands to deliberately choose two-wheel drive over all-wheel drive for its most driver-focused offering.

So the 670-4 SV retained an all-wheel drive system and looked instead to reduce mass by using lightweight materials, especially ultra-high strength sectional steel for the SV's spaceframe chassis, and by eschewing all unnecessary luxuries.

Lamborghini's talented engineers were able to use thinner tubing, thereby saving 20kg from the spaceframe alone, while still increasing its rigidity by 12% over and above its predecessor's spaceframe. Alcantara replaced leather because it is lighter, and carbon fibre was used extensively throughout the car, including for the front and rear bumpers, the lower sill air intake covers, the rear wing and its struts, the third brake light casing, the seat shells, the floor, the transmission tunnel cover, and the door skins. These modifications saved the 670-4 SV 100kg in kerb weight over a comparable LP 640 Coupe – 33kg of this weight saving coming from the chassis and the exterior, 33kg from the drivetrain, and 34kg from the interior.

The 670-4 SV's engine was modified with a revised air intake system, a completely new exhaust system, and an optimised valve train, and with these changes the car produced 661bhp at 8,000rpm, and 660Nm torque at 6,500rpm. Standard on the SV were front axle lift, carbon-ceramic brakes, and an updated E-Gear transmission system with Corsa (Race) and Low Adherence modes. The customer could, however, specify the classic, open-gate, six-speed manual transmission system as a no-cost option.

Sant'Agata also sought to maximise the 670-4 SV's performance by optimising the car's aerodynamics. The SV gained a new and even more prominent front spoiler, a fixed carbon fibre rear wing supported by two prominent carbon fibre struts, a large, full-length, double-decker rear diffuser, and a substantially modified engine compartment bonnet with a transparent centre made from three tiers of hexagonal polycarbonate polymer plates. A truly massive rear wing called the Aeropack Wing was also available as a very expensive optional extra.

The SV's interior was largely swathed in Alcantara and it was dripping in carbon fibre. Sport bucket seats with a carbon fibre shell came as standard, and they featured body-coloured Y-shaped stitching right down the seat centres. SV branding was scattered throughout the cockpit, and this was mirrored on the outside with large SV logos splashed right across the car's rear flanks. The SV was available in one of five standard colours: Arancio Atlas, Giallo Orion, Bianco Isis, Grigo Telesto, and Nero Aldebaran. Extra-cost optional colours included Bianco Canopus and Nero Nemesis. At launch a standard LP 670-4 SV cost £221,335 in the United Kingdom – an 18% premium over a standard LP 640 Coupe.

This swansong of the Murcielago range was powered by the final and ultimate iteration of the classic Bizzarrini-derived Lamborghini V12 engine, which could propel it from standstill to 60mph in 3.2 seconds and on to a top speed of 209mph with the larger Aeropack rear wing fitted, or to 212mph with the standard rear wing.

THE REVENTÓN COUPE

The Reventón was more than an LP 640 in a startlingly modern and beautiful ballgown. Despite being almost identical mechanically to the LP 640, this highly limited edition car – available in strictly controlled numbers in both coupe and roadster versions – was a pathfinder for Lamborghini in many different ways. Firstly, the Reventón was outrageously expensive, allowing Lamborghini to test if it could penetrate this very niche market segment. Secondly, it showcased Sant'Agata's in-house design talent. Thirdly, it was a design trailblazer for Lamborghini's future cars. Fourthly, through being the first production Lamborghini to go directly from computer-aided design (CAD) sketches to production it acted as a test bed for Sant'Agata to examine how it could optimise the process of transferring ideas from the concept and design stage to the production phase. Finally, it acted as a test medium to see how efficiently Lamborghini's Centro Stile could collaborate with Lamborghini's Research and Development Department – and to identify any chokepoints.

The Reventón Coupe was launched at the 2007 Frankfurt Motor Show, and Sant'Agata limited production to just twenty cars. By this time Centro Stile's Head of Design was Manfred Fitzgerald and he was happy to openly admit that the Reventón was an entirely design-led project. Lamborghini's own market research had shown that Sant'Agata had a very affluent and loyal customer base who wanted true exclusivity and were happy to pay for it. Lamborghini decided to exploit this sales and marketing opportunity by designing a car with truly flamboyant interior and exterior aesthetics as its key selling point, while using the LP 640's tried and tested drivetrain as the Reventón's mechanical platform.

The Reventón's engine was a blueprinted version of the LP 640's, and an unaltered E-Gear robotised manual transmission system came as standard. Through careful porting, polishing, and balancing of the engine's internals, the car's power was increased by 10bhp to 641bhp at 8,000rpm. With a peak torque of 660Nm at 6,000rpm, the 1,665kg E-Gear equipped Reventón coupe could accelerate from rest to 100km/h in 3.40 seconds and reach a top speed of 211mph. So nothing particularly

Twins – orange, mid-engined, naturally aspirated, 2-seater convertibles with flank air-intakes and single wiper-blades. Romano Artioli who once owned Lotus, and whose grand-daughter Elisa gifted her name to the Elise, was great friends with Ferruccio Lamborghini and worked closely with Paolo Stanzani on the Bugatti EB110.

special on the mechanical or performance front, and certainly nothing here to justify a price tag of 1 million Euros (before local taxes), which would have bought about five LP 640 coupes.

No, it was the Reventón's body panels and its cabin furniture, particularly its dashboard, that would allow it to command such a massive premium. The Reventón was based on late twentieth-century jet fighter aesthetics, and in particular the styling of the F-22 Raptor. Fitzgerald took his entire Centro Stile team to a United States Air Force base in northern Italy to examine a variety of fighter jets, and the inspiration the team drew from the F-22 would lead them to design the Reventón's interior and exterior. Alessandro Serra led on the design of the Reventón's exterior body panels, which were largely made of carbon fibre.

The Reventón was highly angular and took the triangle and the quadrilateral as its design motifs. It had a distinctly nose-forward appearance thanks to a new, sharply protruding, front bumper-cum air dam, which housed two massive, quadrilateral air inlets. This gave the Reventón's frontal aspect a very aggressive appearance. The front wings were completely redesigned to accommodate the new tapering quadrilateral headlights, and the guillotine doors and the rear flank body panels were sharply creased in origami-like fashion. The rear bodywork was equally as aggressive in appearance with a pointed lower bumper, huge angular rear air vents, a single six-sided exhaust pipe, a redesigned and transparent engine compartment bonnet, and a prominent rear diffuser.

The Reventón featured a dramatic new wheel design boasting five Y-pattern spokes with opaque carbon fibre fins screwed into each of these black aluminium spokes. These fins create a wind turbine effect, drawing in air to cool the standard carbon ceramic brake discs hidden behind the spokes. Light-emitting diodes and Bi-Xenon bulbs now provided front and rear lighting.

The stand-out feature of the Reventón's interior was its dash display instrumentation, which took inspiration from, and mirrored the appearance of, the latest fighter jets' instrument panels. Three Digitech Thin Film Treatment (TFT) liquid crystal displays were encased within an angular aluminium milled binnacle, which itself was encased within a carbon fibre outer casing. At the touch of a button the driver could choose to have the car's various parameters displayed either in quasi-analogue form on circular dials, or in a much more radical display utilising two converging lines simulating a pilot's view of a landing strip with the various performance and check parameters displayed as illuminated bars.

Every one of the twenty Reventón Coupes that left the factory was painted in a new colour specially developed for this car: matt military grey-green with metallic flecks. Sant'Agata was as unadventurous in naming this new paint – which they simply called Reventón – as they were adventurous in designing the Reventón's aesthetics. The Reventón coupe could sprint from rest to 100km/h in 3.4 seconds en route to 211mph, and the entire build run was sold out prior to production even starting.

THE REVENTÓN ROADSTER

The Reventón Roadster was launched at the 2009 Frankfurt Motor Show. It featured the same unique and startling interior and exterior aesthetics as its hard-top counterpart but was even more exclusive and expensive. Only fifteen Roadster units were produced and each was priced at 1.1 million Euros (before local Taxes), making the Roadster 25% more exclusive and 10% more expensive than its Coupe sibling. The Roadster used the same powerplant as the LP 670-4 SV, boasting 661bhp at 8,000rpm and 660Nm torque at 6,000rpm. At 1,690kg it weighed 25kg more than its coupe counterpart and was also less aerodynamic, which explains why its top speed was only 205mph. Despite all the above, Lamborghini claimed that the Reventón Roadster could sprint from rest to 100km/h in the same 3.4 seconds as the Coupe. A new matt grey paint with metallic flecks called Reventón Grey was developed specially for the Reventón Roadster.

THE SPECIALS

In 2003 Lamborghini presented itself with a 40th birthday present in the form of the limited edition Murcielago 40th Anniversary. Only fifty units were produced and all were painted in Verde Artemis with contrasting anthracite coloured wheels. Fortieth Anniversary logos graced the car's rear flanks and cabin, but it was mechanically identical to the standard car save for a modified exhaust system.

The Murcielago 640 Versace edition coupe was the result of a collaboration between the Gianni Versace fashion house and Sant'Agata, and this limited edition variant made its debut at the 2006 Paris Motor Show. Twenty units were built, ten in Isis white and ten in Nero Aldebaran black. All sat on high-gloss black Hermera wheels, and the cabin was swathed in Nappa leather embossed with Versace's signature Greek Fret motif. Somewhere between one and four 640 Versace Roadsters were also built.

The last open-top Murcielago that Sant'Agata built was the Murcielago LP 650-4 Roadster, which was released in 2009. All fifty cars were finished in Grigio Telesto with Arancio Atlas detailing on the VACS vents, side sills, brake callipers, and front spoiler. These cars boasted an additional 10bhp over the standard LP 640 Roadsters and hence their alphanumeric designation (650 CV or 641bhp or 478 kW). Lamborghini claimed a top speed of 205mph and a 0–100km/h time of 3.40 seconds.

Exclusive to the Chinese market, and limited to just ten units, the Murcielago LP 670-4 China Limited Edition was launched at the 2010 Beijing Motor Show. Mechanically identical to the standard LP 670-4 SV, every car was painted in flat grey with a central orange strip sandwiched by black borders, while large SV logos adorned the flanks. Arancio detailing also dominated the Alcantara-clad cabin.

The Murcielago R-GT was developed to FIA and ALMS specifications in conjunction with Audi Sport and Reiter Engineering. Designed to compete in the 1,100kg and below class, this rear-wheel drive racing car was powered by a reduced-capacity 5,998cc engine and further hampered with air restrictors. The R-GT's focus on track-work was emphasised by its lowered ride height, its massive fixed rear wing, and its gearing, which could be altered to suit individual race circuits. The Murcielago R-GT's top speed and acceleration were therefore determined by the gear ratios chosen.

CHAPTER 6

RANDOM MUSINGS OF A DILETTANTE

Long-term ownership of a Countach and a Murcielago, and some hands-on experience of several Diablos, has divulged to your authors just how similar and yet dissimilar these three cars are. That they are consonant should come as no surprise as they were all designed and built by a small group of people working to the same principles and to comparable briefs, and whose tenures at Sant'Agata often overlapped. Equally, their dissonance should not be a thunderbolt out of the blue – thirty-nine years of automotive evolution will always bring some divergence.

Marcello Gandini's genius remains showcased by the fact that his extreme wedge-shape styling and his revolutionary guillotine door design lasted the full span of the Holy Trinity's long reign, and then some. In fact they remain relevant to Lamborghini's 2025 flagship, and to many other current top-tier supercars produced by rival manufacturers – albeit in modified form. Likewise, the Holy Trinity's earliest powertrain details, drivetrain configuration, and chassis construction – as seen in the very first production Countach's Bizzarrini V12, Stanzani's longitudinal placement of this engine and his radical South-North orientation of the engine-gearbox, and Marchesi's inspired lattice-like spaceframe chassis – have all stood the test of time, although they too have had to be altered and fettled to make them relevant for each consecutive time period, and for each successive generation.

In fact, just like how the steam-powered vehicle that the Flemish missionary and astronomer Ferdinand Verbiest built in 1672 for the Kangxi Emperor of China evolved into Nicolas-Joseph Cugnot's steam-powered, artillery-lugging, tricycle of 1769, and then transmuted into Carl Benz's three-wheeled Benz Patent-Motorwagen Nummer 1 of 1886, Sant'Agata also progressively evolved its Holy Trinity. Lamborghini recognised that its supercars could not resist the march of progress any more than Verbiest's boiler-endowed contraption, and the factory therefore followed the well-trodden engineering path of adapting to contemporary demands and exploiting newly available technology. All to ensure that Sant'Agata's offerings remained the fittest, and therefore deserved to survive.

The Countach was indisputably the fittest of its generation, while the Diablo and the Murcielago were among the fittest of their time – the two consecutive periods after 1990 throwing up many more rivals than the Countach ever had to compete against. There may be no better way to explain the common specialness that these three supercars share, and the unique disparateness of each of these three exceptional and

Purchasing any of Lamborghini's Holy Trinity is an emotion-driven affair, but make sure that the heart doesn't totally overrule the head. Caveat emptor.

extraordinary vehicles, than by elaborating on how to buy, maintain and enjoy them, and by describing how they drive, and what they are like to pilot.

BUYING

The Countach, Diablo, and Murcielago are all rare cars. Early Lamborghini production numbers remain a grey area, but authoritative sources state that in the period between 1974 and 2010 only 1,998 Countachs, 2,903 Diablos, and 4,099 Murcielagos were built. In stark contrast, Sant'Agata produced 6,087 V8-engined Urus SUVs and 3,962 V10-engined Huracans, in 2023 alone.

So finding a Holy Trinity car that is to one's taste, and more importantly is fit for purpose, is a mammoth task. It took your authors almost three years of intensive searching worldwide to find their 88 ½ Rosso Siviglia Countach Chassis JLA12399. We wanted a one-owner car in excellent mechanical and cosmetic condition with a comprehensive service record. Not a particularly onerous set of requests one would have thought. However, despite not demanding a particular interior or exterior colour, or the presence or absence of optional addenda like a rear wing, numerous specialist vendors threw up their hands in horror, saying that this was an almost impossible list of wants. The most memorable and hilarious response came from a European dealer, who said that it might just be possible for him to find us a one-owner Countach, but that the car would have to be imported from the Middle East, and that we should expect to find desert sand embedded not only in the floor mats but also within the cylinder bores! We finally found exactly what we wanted on 26 May 2001 from a source much closer to home – from H.R. Owen in London – at that time Lamborghini's only official franchise in the United Kingdom. But it had taken us years of scouring the market.

Finding a suitable Murcielago was not easy either. We wanted an Arancio Atlas Murcielago Roadster in excellent condition with a full service history, and preferably an early 6.2 version as we prized Luc Donkerwolke's uncluttered, more purist, Bauhaus approach over the better-performing LP 640's more glitzy and blingy aesthetics. We started looking in 2013 and it took us nineteen months to find Chassis 1564. Patience is a virtue, nay a necessity, when buying one of Lamborghini's Holy Trinity – otherwise you risk picking up someone else's trouble, and paying handsomely for it to boot.

Regardless of model or model variant, purchasing any of the Holy Trinity is akin to entering a viper's den. Our purchase of Countach Chassis JLA12399 through H.R. Owen was an entirely pain-free delight, and we still remember dealing with Jason Barker, then Lamborghini UK's sole salesman, fondly. In contrast, we bought our Murcielago Roadster from a so-called 'Lamborghini specialist' in the south of England and this purchase was a nightmare that ended up with us taking the vendor to court – the first time we had ever appeared in front of a judge. The trader ended up paying us thousands of pounds in damages and for the repair of a gearbox oil leak that was present on delivery. It would have been infinitely quicker, cheaper, and more pleasant for all concerned had this Del Boy of supercar purveyors immediately acknowledged and repaired the simple leak, rather than try to wriggle out of his legally mandated responsibilities. Thankfully this is where United Kingdom consumer law steps in to protect the buyer, and for readers from other countries, you too will almost certainly have similar consumer protection legislation in place to fall back upon should the need arise.

Ultimately though, caveat emptor. The best thing that a potential buyer can do to protect himself from a dud purchase is to have a proper Pre-Purchase Inspection (PPI) carried out by an acknowledged and

Gandini's extreme wedge shape and trapezoidal design language married to Bizzarrini's stalwart V12 and a lattice spaceframe chassis underpins every Holy Trinity model.

Legendary, or more accurately infamous, Italian supercar fragility means that water-tight garaging is a necessity and not a luxury.

experienced Lamborghini engineer. However, even this is not foolproof. Wet behind the ears, we bought our Countach without a prior PPI and all turned out well, while we paid out for a full and thorough PPI at a Lamborghini Main Dealer before agreeing to buy our Murcielago and still ran into the difficulties detailed above.

Remember though that these are rare cars and finding one to your exact specification might well be difficult, so you might have to compromise and deal with shady characters whom you might not otherwise choose to do business with. In our case, we have not seen another Murcielago Roadster to our spec come up for sale in the ten years since we bought ours. So hold your nose and deal with the shifty and slippery if you absolutely have to, but take every precaution before handing over your hard-earned money. At the end of the day you are buying the car and not the vendor.

Due diligence will involve checking the paperwork carefully and then investigating further. Scrutinise with a fine toothcomb the car's factory specification sheet, original purchase invoice, MOT certificates, and service history booklet. Speak to previous owners, interrogate the technicians who have worked on the car, question Lamborghini Club members who might know the car or its previous owners, and pay for an external specialist firm to see if there is any outstanding finance on the car or if it has been involved in an accident.

Check also if the car has provenance – Countach Chassis JLA12399 had featured extensively in Crowood's 1990 Countach book by Peter Dron, and also had the starring role in a 1989 Alpine-Hermès-Cartier television commercial. Similarly, a photograph of Murcielago Roadster Chassis 1564 taken at Goodwood in 2008 appears in Wikipedia's Murcielago article, and both cars are now Beaulieu National Motor Museum photographic reference examples with their images appearing in that institution's Picture Library. Provenance adds a soupçon of spice to ownership.

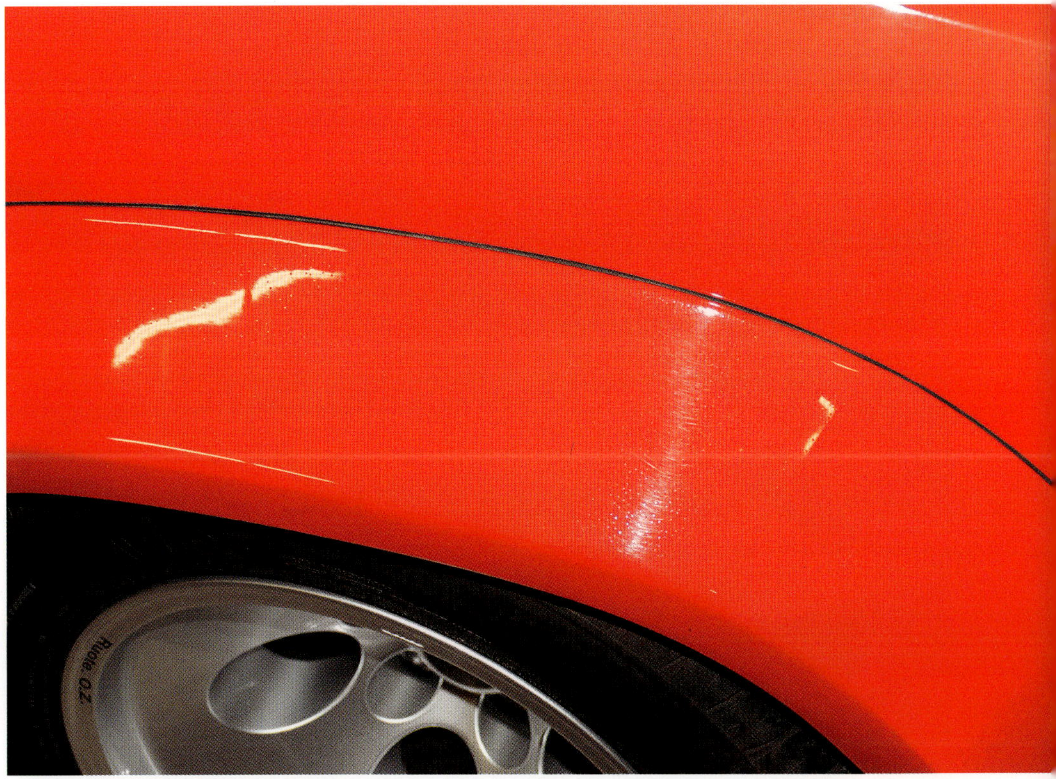

Zero-restoration Countachs are becoming very, very rare. Why desecrate provenance and erase history for mere artificial beauty? These wheelarch micro-blisters were present at delivery in 1988 and have not altered since. Reason: Sant' Agata was still learning, and struggling, to paint composites at that point in time.

MAINTENANCE

The Countach, Diablo, and Murcielago are hand-built cars and no two cars are exactly the same. But one thing that all the Holy Trinity cars have in common is that they need, and deserve, more care and attention than the average car.

Lamborghini's Holy Trinity need expert fettling from experienced mechanics that know each model's idiosyncrasies and peculiarities.

The most humble looking of garages might house the most experienced and dedicated of mechanics. Mike Pullen's workshop is a case in point.

They need to be serviced by experts who know their individual idiosyncrasies intimately – a Countach expert may not have the necessary expertise to work on a Murcielago – and this costs. For the last twenty-four years our Countach has been attended to by Mike Pullen, which entails a 320-mile round trip in an easterly direction, while our Murcielago has most recently been worked on by David Jenkins, the Master Technician at Lamborghini Pangbourne, which requires a 200-mile journey in a northerly direction. This is just part of the cost of vintage Lamborghini ownership – the price is more than simply financial.

Lamborghini's Holy Trinity all boast a steel spaceframe chassis that is prone to rust, particularly if exposed to road salt. For this reason alone we put our Countach and Murcielago into partial hibernation between November and April. But religiously, every six weeks, we start up the cars and drive them up and down our driveway, which is never salted. This is time-consuming as the engine and gearbox have to be brought up to full operating temperature. For every 1 litre of petrol combusted within a Bizzarrini V12 engine, 1 litre of water is produced, and this water will remain within the drivetrain's lubricants, compromising their function and effectiveness, unless the drivetrain's temperature is sufficiently high to evaporate away this water. So we never just start up and idle a car for five minutes – this will probably do more harm than good. Instead the cars' engines need to kept running at least until the radiator fans kick in – which in mid-winter can take twenty minutes with the Murcielago and up to forty-five minutes with the Countach. Driving the cars, even small distances, helps exercise the clutch, brake, and accelerator mechanisms, and works the gearbox, while also preventing flat-spotting of the wheel bearings and tyres.

The Holy Trinity, particularly the Countach, have typical Italian auto electrics, which are just as susceptible to damp and corrosion as the steel spaceframe chassis. This is another reason why these cars need to be kept in a dry, preferably temperature and humidity controlled, garage. For long-term storage in cold, damp, rainy England, a dehumidifier and a battery tender are necessities.

These three cars are now quite valuable, and it is advisable to store them in a secure, alarmed garage, sheltered behind different and varied layers of protection and remote surveillance, including heavy-duty retractable bollards, closed-circuit television cameras and satellite tracking systems.

Restoration versus patina is a much discussed and debated topic. We abhor Pebble Beach-style restorations – why not wear the outward signs of a life well lived with pride? This applies to the owners as much as to the Lamborghinis!

While we choose to reject unnecessary cosmetic surgery for our Sant'Agata steeds, we certainly believe that any accident damage, or corrosion, or mechanical malfunctions, or necessary drivetrain maintenance, should always be carried out in a timely and comprehensive manner.

Zero restoration classic supercars, meaning supercars that have never been subjected to unnecessary cosmetic improvements just to make them look new and shiny, are an increasingly rare breed – in the case of the Countach, and increasingly the Diablo, we might venture to say an almost extinct breed. True connoisseurs are beginning to appreciate these sometimes less than cosmetically perfect cars more and more. Unnecessary restoration hides, and sometimes extinguishes completely, the telltale signs of how a car was constructed in period. Unneeded restoration therefore steals history, and this is why we are so against it.

DRIVING

Even after a quarter of a century, simply walking towards our parked-up Countach always sets our hearts aflutter. On those occasions when we know that we are going to be starting her up and taking her for a drive the excitement levels reach for the stars. Our Murcielago performs this very same trick equally effortlessly, as have the few Diablos that we have been lucky enough to drive. We will limit ourselves to talking about driving our late Countach and our early Murcielago as these are the cars that we have the most experience with, and by happenchance they neatly sandwich, chronologically speaking, the Diablo.

Starting up the Countach with her sextet of Weber carburettors, requires turning the ignition key to its first stop, which activates the fuel pumps, and then slowly and deliberately pumping the accelerator pedal six times to prime the Webers. Following which the clutch pedal needs to be fully depressed before twisting the ignition key all the way to its final stop. Meanwhile, you, the driver, should be praying. The starter motor may or may not whirr, and if it does the Bizzarrini V12 may or may not catch. And if the engine does catch, it may or may not run for more than a couple

Essential maintenance on a Countach should never be delayed or skimped upon.

Above: A sextet of down-draught Weber 44 DCNF carburettors from an 88 ½ Countach undergoing routine refurbishment in 2011, after 23 years of sterling service.

Opposite: A seldom-seen view of a Murcielago's V12 engine.

of seconds before stuttering to a silent stop. Once the V12 starts running it stumbles lazily into life, first on just a few cylinders and finally, after about twenty seconds, on the full dozen. And even then it idles lumpily for a further half minute, albeit getting ever smoother with each passing second. The clutch pedal must now be let out slowly and progressively otherwise the engine might, or might not, stall. You will now understand why prayer is a fundamental and irrevocable part of Countach ownership.

Starting up the Murcielago Roadster is just as involved. First press the immobiliser off button on the key fob, and then insert the key into its barrel and twist it to the first stop. The car will bleep, screech, and hiss at you, and thirty-four multi-coloured warning lights will flash up on the dash display. Again prayer and patience are needed. Pray hard that all these lights will disappear quickly, and wait patiently for them to do so, as any that remain illuminated (other than the oil pressure and battery warning lights) portend of a headache-inducing and expensive trip to the mechanic. Once the thirty-two remaining lights have all disappeared, turn the key fully while keeping both feet off all the pedals, and if God has been listening, the starter motor's brief whirr will be followed almost instantly by the V12 engine catching. The Murcielago's V12 settles to a smooth idle within a few seconds, and no other driver gymnastics are usually necessary hereafter. So similarities and differences in the starting procedures – just as you would expect when comparing and contrasting a 1988 vintage car with its 2005 classic grandchild.

But, of course, before you can start up either of these cars you first have to get into their cockpits – which demands acrobatic manoeuvring through those narrow guillotine door apertures, and then across those wide, high, sills. Once seated in either the Countach's hammock seat or the Murcielago's more conventional slab seat, the driver will have excellent forwards visibility but no clue as to where the low nose ends. Side visibility is middling to poor, while rear visibility is appalling,

The disassembled top section of a Murcielago's V12 engine.

especially in the later cars, which boast a mille-feuille-shaped hump in the middle of the engine compartment bonnet.

The Countach, Diablo, and Murcielago all hold vast volumes of engine oil and transmission oil, and it is absolutely vital for the long-term health of their drivetrains that none of these cars' engines are revved until these lubricants have warmed up and reached their optimal operating temperatures. Which, depending on the ambient air temperature, can take up to twenty-five minutes, or about fifteen miles of gentle driving. Do not be fooled by the coolant temperature gauge – whose needle will sprint into the vertical position relatively quickly. It is the engine oil temperature gauge that is the critical dial to watch.

So, both these cars, once the fastest and fastest-accelerating cars of their respective generations, are also, in the hands of a mechanically sympathetic owner, amongst the slowest cars out in the real world!

And if you were expecting jaw-dropping, low-speed acceleration, the Countach in particular is disappointing. Yes, its Bizzarrini engine throbs to a promising and distinctive martial beat even at idle, and this loud, dull Napoleonic thud becomes ever more metallic and shrill once past 1,500rpm. But the car's acceleration at this point in its rev range is at odds with its fearsome reputation. Those contemporary road testers must have been absolutely brutal with the revs and the clutch to obtain a 0–60mph time of 4.9 seconds. No half-sane owner would subject their grand old lady to such abuse today.

No, the Countach comes into its own at higher engine and road speeds. Once past 4,500rpm and 70mph the whole drivetrain becomes ever more smooth and ever less agricultural. And once into three figures the Countach just takes on a totally different character – everything now gels, everything now flows. By this stage the accelerator pedal is past its sticky, gnarly, first few centimetres, the gear lever slips ever more fluidly into its next slot with increasing revs, the Countach's infamous 1970s-era brakes retard the car better once the pads and discs have some genuine heat in them, and the previously slightly heavy steering lightens up just enough to become ideally weighted.

At speeds where top-tier sports cars hit their aerodynamic brick wall, the notoriously un-aerodynamic Countach just keeps on accelerating. And it does so with a basso profundo voice and in a deeply visceral style – the engine screaming, the gearbox howling, the hunkered down suspension providing a miraculously limousine-like ride on smooth tarmac, and the steering wheel writhing gently while accurately and faithfully transmitting every nuance of the road surface to the driver's fingers and palms.

Once, a long time ago, somewhere in Europe, we took our then 13-year-old Countach up to about 7,800rpm in fifth gear, and it was an other-worldly experience. This was such a life-affirming, never to be forgotten, event that we devoted the opening chapter of our Countach book to the minute or so that it took JLA 12399 to reach this engine speed in this gear.

And herein lies the conundrum of living with any of Lamborghini's Holy Trinity cars – you cannot begin to exploit their performance, their raison d'être, on today's heavily trafficked, heavily policed, speed camera-infested roads. The mind-searing drive we referred to above was carried out at 5 A.M. on a deserted, arrow-straight, five-lane per carriageway motorway that we knew very well, on a mid-summer's morning when there wasn't even a hint of dew on the road surface, and just after the car had been thoroughly inspected and freshly serviced. Getting all those stars aligned was pure luck – we had promised each other that the slightest imperfection in the car or the road conditions would mean immediately aborting this long-planned attempt. Today, in southern England, it would be impossible, and totally irresponsible, to repeat anything remotely like this. As already stated, this was a long time ago – in a much more permissive era, and in a much more speed-tolerant country.

All of the above is true for the Murcielago too. We have never attempted to max Chassis 1564, again for the reasons just stated above. But what we can say definitively is that the Murcielago with its four-wheel drive system, its traction control system, and its various other electronic safety nets and refinements, will at least put up a modest show of saving you in extremis, whereas the Countach will happily send you on your way to your Maker.

We choose to drive our Countach and Murcielago very early in the morning, on warm, dry days. The cars are usually back in the garage by 6.45 A.M. It is under these circumstances, and in these conditions, that we find that we can best exploit and enjoy these cars. The Countach and Murcielago thrive on high engine speeds, and are not at ease sharing tarmac with other moving vehicles. Just as we are not comfortable being in the vicinity of other curious drivers. So, in our opinion, an alarm clock is an essential and indispensable requirement for fully enjoying any of the Holy Trinity cars.

Ultimately, the Holy Trinity cars transcend mere vehicular transport. They are truly extraordinary vehicles, although herein lies both their forte and their foible. Hopefully, any owner, however long their possessorship, will only be a temporary custodian of any one of these

A totally visceral 1971-era supercar with manual steering, minimally assisted brakes, two-wheel drive, a heavy clutch pedal and a recalcitrant gearchange (1988 Countach 88½ QV).

three supercars. The Countach, Diablo, and Murcielago deserve to last beyond the biblical three score and ten years afforded to humans. Yes, they are cars, but they are also much, much more. They are works-of-art and historical artefacts, and they represent a rare and long-lived lineage. A bloodline that tells the story about the zenith of the internal combustion engine and motoring itself – the twin technological advances that liberated hundreds of millions to more easily enjoy their wider surroundings, while also nurturing countless societies, and powering endless economies. Long after the sale of new internal combustion engine cars is outlawed, and later still when petrol is seen as an antique propellant, the Countach and its two immediate descendants will still astound and enthral, if only as static displays in motoring museums. Every one of Lamborghini's Holy Trinity needs to be cherished and preserved for future generations.

Despite power-assisted steering, ABS, four-wheel drive, and a robotised manual transmission system, this 2001-era supercar still sets the heart aflutter every single time (2005 Murcielago Roadster).

Gandini's and Stanzani's Countach was a truly revolutionary car. Automotive evolution, driven by almost four decades of technological innovations and irrepressible socio-politico-economic factors, begat the derivative but still spectacular Diablo and Murcielago.

BIBLIOGRAPHY

Lamborghini Countach, Jean-Francois Marchet & Peter Coltrin (Osprey, Autohistory Series, 1986)

Lamborghini Countach: The Complete Story, Peter Dron (Crowood, Autoclassics Series, 1990)

Lamborghini Diablo, Stefano Pasini (Automobilia, 1991)

Lamborghini Murcielago, Pathmanathan (Veloce, 2019)

Lamborghini Countach, Pathmanathan & Reck (Evro, 2021)

Lamborghini Diablo, Pathmanathan & Reck (Crowood, 2023)

Countach Specification Table

	LP 500 Prototype	LP400	LP400 S	LP500 S	5000 QV	88 1/2 QV	25th Anniversary
Production Dates	1971	1974–78	1978–82	1982–85	1985–87	1987–88	1988–90
Engine capacity	4,971cc	3,929cc	3,929cc	4,754cc	5,167cc	5,167cc	5,167cc
Bore	85mm	82mm	82mm	85,5mm	85,5mm	85,5mm	85,5mm
Stroke	73mm	62mm	62mm	69mm	75mm	75mm	75mm
Compression ratio	10.5:1	10.5:1	10.5:1	9.2:1	9.5:1	9.5:1	9.5:1
Valves per cylinder	2	2	2	2	4	4	4
Ignition system	Twin coils/ distributors	Twin coils/ distributors	Twin coils/ distributors	Electronic	Electronic	Electronic	Electronic
Fuel Provision	Weber 45DCOE	Weber 45DCOE	Weber 45DCOE	Weber 45DCOE	Weber 44DCNF	Weber 44DCNF	Weber 44DCNF
Transmission	5-speed manual	5-speed manual	5-speed manual	5-speed manual	5-speed manual	5-speed manual	5-speed manual
Steering	unassisted rack and pinion	unassisted rack and pinion	unassisted rack and pinion	unassisted rack and pinion	unassisted rack and pinion	unassisted rack and pinion	unassisted rack and pinion
Suspension	wishbone, coil spring, ABR	wishbone, coil spring, ABR	wishbone, coil spring, ABR	wishbone, coil spring, ABR	wishbone, coil spring, ABR	wishbone, coil spring, ABR	wishbone, coil spring, ABR
Drive	rear wheel drive	rear wheel drive	rear wheel drive	rear wheel drive	rear wheel drive	rear wheel drive	rear wheel drive
Brakes	steel ventilated discs	steel ventilated discs	steel ventilated discs	steel ventilated discs	steel ventilated discs	steel ventilated discs	steel ventilated discs
Wheel size (front)	7.5J × 14	7.5J × 14	8.5J × 15	8.5J × 15	8.5J × 15	8.5J × 15	8.5J × 15
Wheel size (rear)	9J × 14	9J × 14	12J × 15	12J × 15	12J × 15	12J × 15	12J × 15
Tyre size (front)	205/70 VR 14	205/70 VR 14	205/50 VR 15	205/50 VR 15	225/50 VR 15	225/50 VR 15	225/50 VR 15
Tyre size (rear)	215/70 VR14	215/70 VR14	345/35 VR 15	345/35 VR 15	345/35 VR 15	345/35 VR 15	345/35 VR 15
Peak power	440bhp@7,400rpm	375bhp@8,000rpm	353bhp@7,500rpm	375bhp@7,000rpm	455bhp@7,000rpm	455bhp@7,000rpm	455bhp@7,000rpm
Peak torque	366lb-ft@5,000rpm	266lb-ft@5,500rpm	267lb-ft@5.500rpm	303lb-ft@4,500rpm	369lb-ft@5,200rpm	369lb-ft@5,200rpm	369lb-ft@5,200rpm
Kerb weight	1,322kg	1,300kg	1,371kg	1,480kg	1,488kg	1,488kg	1,490kg
Overall length	4,140mm	4,140mm	4,140mm	4,140mm	4,140mm	4,140mm	4,140mm
Wheelbase	2,450mm	2,450mm	2,450mm	2,450mm	2,450mm	2,450mm	2,450mm
Overall width	1890mm	1890mm	2000mm	2000mm	2000mm	2000mm	2000mm
Overall height	1,030mm	1,070mm	1,070mm	1,070mm	1,070mm	1,070mm	1,070mm
0–100km/h (0–62mph)	unknown	5.6s	5.9s	5.6s	4.9s	4.9s	4.9s
Top speed	186mph	196mph	181mph	180mph	183mph	183mph	183mph

Diablo Specification Table

	Diablo	VT	SE30	SV	Roadster	2nd Gen MY 1999 VT	2nd Gen MY Roadster	2nd Gen MY SV	GT	6.0 VT
Production Dates	1990–1998	1993–1999	1993–1994	1995–1999	1995–1999	1999–2000	1999–2000	1999–2000	1999–2000	2000–2001
Engine capacity	5707cc	5707cc	5707cc	5707cc	5707cc	5707cc	5707cc	5707cc	5992cc	5992cc
Bore	87mm	87mm	87mm	87mm	87mm	87mm	87mm	87mm	87mm	87mm
Stroke	80mm	80mm	80mm	80mm	80mm	80mm	80mm	80mm	84mm	84mm
Compression ratio	10	10	10	10	10	10	10	10	10.7	10.7
Valves per cylinder	4	4	4	4	4	4	4	4	4	4
Ignition system	Electronic	Electronic	Electronic	Electronic	Electronic	Electronic	Electronic	Electronic	Electronic	Electronic
Fuel Provision	Fuel injection	Fuel injection	Fuel injection	Fuel injection	Fuel injection	Fuel injection	Fuel injection	Fuel injection	Fuel injection	Fuel injection
Transmission	5-speed manual	5-speed manual	5-speed manual	5-speed manual	5-speed manual	5-speed manual	5-speed manual	5-speed manual	5-speed manual	5-speed manual
Steering	Manual Rack +Pinion	Power assisted Rack +Pinion	Manual Rack +Pinion	Power assisted Rack +Pinion	Power assisted Rack +Pinion	Power assisted Rack +Pinion	Power assisted Rack +Pinion	Power assisted Rack +Pinion	Power assisted Rack +Pinion	Power assisted Rack +Pinion
Suspension	wishbone, coil spring, ABR	wishbone, coil spring, ABR	wishbone, coil spring, ABR	wishbone, coil spring, ABR	wishbone, coil spring, ABR	wishbone, coil spring, ABR	wishbone, coil spring, ABR	wishbone, coil spring, ABR	wishbone, coil spring, ABR	wishbone, coil spring, ABR
Drive	Rear Wheel Drive	Four Wheel Drive	Rear Wheel Drive	Rear Wheel Drive	Four Wheel Drive	Four Wheel Drive	Four Wheel Drive	Rear Wheel Drive	Rear Wheel Drive	Four Wheel Drive
Brakes	steel ventilated discs	steel ventilated discs	steel ventilated discs	steel ventilated discs	steel ventilated discs	steel ventilated discs	steel ventilated discs	steel ventilated discs	steel ventilated discs	steel ventilated discs
Wheel size (front)	8.5 × 17	8.5 × 17	8.5 × 17	8.5 × 17	8.5 × 17	8.5 × 18	8.5 × 18	8.5 × 18	8.5 × 18	8.5 × 18
Wheel size (rear)	13 × 17	13 × 17	13 × 18	13 × 18	13 × 17	13 × 18	13 × 18	13 × 18	13 × 18	13 × 18
Tyre size (front)	245/40ZR17	245/40ZR17	235/40ZR17	235/40ZR17	235/40ZR17	235/35ZR18	235/35ZR18	235/35ZR18	245/35ZR18	235/35ZR18
Tyre size (rear)	335/35ZR17	335/35ZR17	335/30ZR18	335/35ZR18	335/35ZR17	335/30ZR18	335/30ZR18	335/30ZR18	335/30ZR18	335/30ZR18

	Diablo	VT	SE30	SV	Roadster	2nd Gen MY 1999 VT	2nd Gen MY Roadster	2nd Gen MY SV	GT	6.0 VT
Peak power	492bhp @6800rpm	492bhp @6800rpm	520bhp @7100rpm	510bhp @7100rpm	492bhp @7000rpm	530bhp @7100rpm	530bhp @7100rpm	530bhp @7100rpm	575bhp @7300rpm	550bhp @7100rpm
Peak torque	580 Nm @5200rpm	580 Nm @5200rpm	580Nm @5900rpm	580Nm @5900rpm	580 Nm @5200rpm	605 Nm @5500rpm	605 Nm @5500rpm	605 Nm @5500rpm	630 Nm @5500rpm	620 Nm @5500rpm
Kerb weight	1576kg	1623kg	1449kg	1530kg	1625kg	1625kg	1625kg	1530kg	1490kg	1625kg
Overall length	4460mm	4460mm	4550mm	4470mm	4470mm	4470mm	4470mm	4470mm	4430mm	4470mm
Wheelbase	2650mm	2650mm	2650mm	2650mm	2650mm	2650mm	2650mm	2650mm	2650mm	2650mm
Overall width	2040mm	2040mm	2040mm	2040mm	2040mm	2040mm	2040mm	2040mm	2040mm	2040mm
Overall height	1105mm	1105mm	1105mm	1105mm	1105mm	1115mm	1105mm	1115mm	1115mm	1105mm
0-100km/h (0-62mph)	4.09s	4.09s	3.90s	3.90s	4.00s	3.95s	3.95s	3.85s	3.70s	3.95s
Top speed	328km/h 204mph	325km/h 203mph	333km/h 207mph	328km/h 204mph	328km/h 204mph	335km/h 208mph	335km/h 208mph	320km/h 201mph	338km/h 211mph	335km/h 208mph

Murcielago Specification Table

	6.2 VT	6.2VT Roadster	LP640	LP640 Roadster	LP670-4 SV	Reventon Coupe	Reventon Roadster
Production Dates	2001–2006	2004–2006	2006–2010	2007–2010	2009–2010	2007–2008	2009–2010
Engine capacity	6192cc Dry Sump	6192cc Dry Sump	6496cc Dry Sump	6496cc Dry Sump	6496cc Dry Sump	6496cc Dry Sump	6496cc Dry Sump
Bore	87mm	87mm	88mm	88mm	88mm	88mm	88mm
Stroke	86.6mm	86.6mm	89mm	89mm	89mm	89mm	89mm
Compression ratio	10.7:1	10.7:1	11.0:1	11.0:1	11.1:1	11.1:1	11.1:1
Valves per cylinder	4	4	4	4	4	4	4
Ignition system	Electronic	Electronic	Electronic	Electronic	Electronic	Electronic	Electronic
Fuel Provision	Fuel injection	Fuel injection	Fuel injection	Fuel injection	Fuel injection	Fuel injection	Fuel injection
Transmission	6 Speed Manual or E-Gear robotised	6 Speed Manual or E-Gear robotised	6 Speed Manual or E-Gear robotised	6 Speed Manual or E-Gear robotised	6 Speed Manual or E-Gear robotised	E-Gear robotised	E-Gear robotised
Steering	Power Assisted Rack and Pinion	Power Assisted Rack and Pinion	Power Assisted Rack and Pinion	Power Assisted Rack and Pinion	Power Assisted Rack and Pinion	Power Assisted Rack and Pinion	Power Assisted Rack and Pinion
Suspension	Independent, double wishbones coil springs electronic dampers	Independent, double wishbones coil springs electronic dampers	Independent, double wishbones coil springs electronic dampers	Independent, double wishbones coil springs electronic dampers	Independent, double wishbones coil springs electronic dampers	Independent, double wishbones coil springs electronic dampers	Independent, double wishbones coil springs electronic dampers
Drive	4 Wheel Drive	4 Wheel Drive	4 Wheel Drive	4 Wheel Drive	4 Wheel Drive	4 Wheel Drive	4 Wheel Drive
Brakes	Steel Brakes ABS DRP	Steel Brakes ABS DRP Upgraded	Steel Brakes ABS DRP	Steel Brakes ABS DRP	Carbon-Ceramic standard	Carbon-Ceramic standard	Carbon-Ceramic standard
		8 piston front calipers	Carbon-Ceramic option	Carbon-Ceramic option			
Wheel size (front)	8.5 × 18	8.5 × 18	8.5 × 18	8.5 × 18	8.5 × 18	8.5 × 18	8.5 × 18
Wheel size (rear)	13 × 18	13 × 18	13 × 18	13 × 18	13 × 18	13 × 18	13 × 18
Tyre size (front)	245/35ZR18	245/35ZR18	245/35ZR18	245/35ZR18	245/35ZR18	245/35ZR18	245/35ZR18
Tyre size (rear)	335/30ZR18	335/30ZR18	335/30ZR18	335/30ZR18	335/30ZR18	335/30ZR18	335/30ZR18
Peak power	571bhp @7,500rpm	571bhp @7,500rpm	631bhp @8,000rpm	631bhp @8,000rpm	661bhp @8000rpm	641bhp @8,000rpm	661bhp @8000rpm

	6.2 VT	6.2VT Roadster	LP640	LP640 Roadster	LP670-4 SV	Reventon Coupe	Reventon Roadster
Peak torque	650Nm @5,400rpm	650Nm @6,500rpm	660Nm @6,000rpm	660Nm @6,000rpm	660Nm @6500rpm	660Nm @6,000rpm	660Nm @6,000rpm
Kerb weight	1650kg	1665kg	1665kg	1665kg	1565kg	1665kg	1690kg
Overall length	4580mm	4580mm	4610mm	4580mm	4705mm	4705mm	4705mm
Wheelbase	2665mm	2665mm	2665mm	2665mm	2665mm	2665mm	2665mm
Overall width	2058mm	2058mm	2058mm	2058mm	2058mm	2058mm	2058mm
Overall height	1135mm	1068mm	1135mm	1068mm	1135mm	1132mm	1068mm
0-100km/h (0-62mph)	3.80 sec	3.89 sec	3.4 sec	3.4 sec	3.2 sec	3.4 sec	3.4 sec
Top speed	205mph	199mph	211mph	205mph	212mph	211mph	205mph

INDEX